Fertilizers and Manures

FERTILIZERS AND MANURES

by

Late Sir Alfred Daniel Hall

M.A., Sc.D., LL.D., F.R.S., K.C.B.

Revised by

A.M. Smith

Ph.D., D.Sc., F.R.I.C.

BIOTECH

2025

BIOTECH BOOKS

ISBN 978-81-7622-066-8

Published by: **BIOTECH BOOKS**®
4762-63/23, Ansari Road, Darya Ganj,
New Delhi - 110 002
Phone: 09899295259, 011-43003222
E-mail: biotechbooks@yahoo.co.in
Website: www.biotechbooks.in

Typeset at: **Classic Computer Services**
Delhi - 110 035

PRINTED IN INDIA

DEDICATED TO

SIR CHARLES LAWES-WITTEWRONGE, BARONET

OF ROTHAMSTED

WHO HAS SHOWN IN OTHER FIELDS
THE DISTINCTION AND IMAGINATION
WHICH MARKED HIS FATHER'S WORK
FOR AGRICULTURE

PREFACE TO THE FOURTH EDITION

NEARLY forty years have elapsed since the first edition of this book was published. Based largely upon the renowned Rothamsted experiments whose results constitute an unique set of data obtained over many years of careful experimentation, it was the first comprehensive exposition in English of the principles of fertilizer practice. Sir Daniel Hall was always at pains to discuss the principles involved rather than to translate the figures to other soil types, or to other environments for, until comparatively recently, there have been few other experimental results available for discussion. It is a matter for great regret that he was unable to carry out his intention of preparing a new edition of the book, for the outlook on fertilizers had changed, even since his last revision in 1929, due to the notable increase in the study and production and use of fertilizers.

The first World War was responsible for rapid developments in the fixation of atmospheric nitrogen. The more recent conflict has seen great progress in the exploitation of potassium resources, especially in those parts of the world previously dependent upon German supplies; noteworthy progress has also been made in new methods of dealing with mineral phosphates so as to render them suitable for use by the plant. Then recent investigations on plant growth factors have revealed the importance of many elements previously disregarded as being essential to the health of the plant.

In undertaking the present revision I have, therefore, found it necessary to replace much of the older material by new in order to keep the book within reasonable dimensions. But every effort has been made to preserve its original design as a text-book for the agricultural student, and I have not changed Sir Daniel Hall's general approach to the subject. The description of the sources of a particular class of fertilizer is followed by a discussion of its use and value. Manufacturing processes have been dealt with rather more fully than before, but not so as to demand of the reader more than a slight knowledge of elementary chemistry. I have assumed that the person who uses fertilizers is probably interested to know in a general way how they are produced but is not likely to be concerned with the technical problems of their manufacture. New data on world production and consumption have been introduced for the sake of showing the relative importance of different fertilizers; since figures for the war years are not generally available, the statistics for 1937–38 have been taken as giving the best possible picture of normal conditions.

A few modifications of chapter titles have been made and sectional headings have been incorporated for convenience in reference. Many parts of the book have been recast in order to allow for the change in emphasis on certain aspects of the subject. For example, in Chapters II, III and IV, an attempt has been made to bring the story of the nitrogenous fertilizers more into line with the present-day position. In the case of the phosphatic compounds, due attention has been given in Chapters V and VI to the advances which have been made in the production of more concentrated fertilizers, and to the questions of phosphate fixation and availability. On the subject of the potassic fertilizers, a considerable amount of new material has been added. Farmyard manure has again been fully dealt with, and some adjustments have been made in Chapters VIII, IX and X so as to incorporate sections on such things as composts, poultry manure, hoof and horn meal and town refuse. Chapter XI has been entirely rewritten to deal with fertilizer requirement, and to present an outline of the trace elements in plant nutrition. Chapter XIV has also been rewritten and now contains a short account of the general principles of modern field experimentation.

Room for the new material has been found by reducing the number of tables, by substituting more general statements for detailed results, and by omitting parts of the text which now appear to be of less importance. Many of the original tables whose contents are not only of great historical value, but serve to demonstrate certain principles as clearly as more recent work, have been left unaltered, as have the illustrations.

The question of nomenclature is a difficult one on account of the usage of so many archaic terms in the fertilizer trade. It is hardly to be expected that the traditional language will be lightly changed for it has been incorporated in the titles of important companies and appears in official documents. In the writer's opinion it would simplify matters considerably and be more logical to discuss all fertilizers in terms of nitrogen, phosphorus and potassium, N, P and K, like boron, copper and so on in the newer literature, and to stop referring to P_2O_5 as phosphoric acid and to potash as if it were an element. Similarly it would be preferable to use the modern chemical names of calcium carbonate for carbonate of lime, of sodium nitrate for nitrate of soda, of ammonium sulphate for sulphate of ammonia and so on. Perhaps a revision of the Fertilizers and Feeding Stuffs Act will embody some of these changes; in the meantime, many chemists have perforce to use the conventional terms to avoid confusion with merchants and farmers.

In my search for reliable up-to-date information, I have been greatly assisted by Dr. Appleby and his colleagues of the Billingham Division of Imperial Chemical Industries Ltd.; by Dr. Cowie, of Potash Ltd.; by Mr. Sandral and Mr. Irvine, of the British Sulphate of Ammonia Federation Ltd.; by Mr. Tubb, of the Nitrate Corporation of Chile Ltd.; by Messrs. Palestine Potash Ltd., all of whom have unreservedly placed material at my disposal. I have also drawn freely upon Gray's valuable compilation on *Phosphates and Superphosphate*. Repeated reference will be found in the text to an article by Crowther and Yates which appeared in the *Empire Journal of Experimental Agriculture* (1941), and whose title, "Fertilizer Policy in Wartime," does not reveal its valuable summary of information collected from many thousands of experiments, or its study on growth curves and fertilizer responses. Finally, I am greatly indebted to many of my colleagues, in particular to Dr. Aitken, Dr. Comrie, Dr. Gibson and Mr. Walker, for advice and criticism on sections of the book.

A. M. SMITH.

PREFACE TO THE FIFTH EDITION

DURING the preparation of the previous edition the latest available statistics on the production and consumption of fertilizers were those compiled for 1938. Since that date there has been a remarkable development in the fertilizer industry to meet agricultural demands in different parts of the world. For example, the output of both nitrogen and potassium has doubled and that of phosphorus increased by 75 per cent in the last 15 years. I have taken the opportunity of bringing that kind of information up to 1953 or later. Prices have also risen and have been adjusted to U.K. 1955 quotations ; it is to be noted that subsidy is no longer included in the price of fertilizers in this country but is recovered by the purchaser. Numerous changes in the text and tables have therefore been necessary, but the general treatment of the subject has not been changed.

A. M. SMITH.

PREFACE TO THE ORIGINAL EDITION

THE use of some form of fertilizer is becoming more and more a mark of modern agriculture. Though many farmers, and among them some of our best, still profess to scorn all artificial manures and pin their faith on the dung made by their stock, they none the less are buying the elements of fertility—nitrogen, phosphoric acid, and potash—in the cakes and other feeding stuffs which they bring from some outside source and consume on their farms. It is the continual introduction of plant food from outside which distinguishes modern intensive methods of cultivation from the old farming. Prior to a period which roughly coincides with the foundation of the Royal Agricultural Society of England in 1838, the farmer, living on the inherent capital of the soil, was forced into a conservative system of cultivation, which by restoring in the dung the greater part of what had been taken from the soil by the crops, would reduce the losses from his land to a point where they would be more or less balanced by the natural recuperative processes at work in the soil. In consequence the level of production was low, and it was the discovery and introduction of artificial fertilizers and feeding stuffs—nitrate of soda, guano, the phosphates, cotton cake, maize, etc.—which enabled the British farmer to raise his output per acre by at least 50 per cent. during the reign of the late Queen. It is true that all intensive farming in the United Kingdom received a great set-back in the 'eighties and 'nineties, when the continued opening up of new areas of virgin soil and the fall in freights filled the country with corn and meat at prices below our cost of production under the conditions then prevailing, because declining prices cannot be met by more intensive methods, but only by a reduction in the expenditure. However, we are steadily recovering from that position: the supply of rich virgin soil is not without a limit nor are its riches inexhaustible; the cost of production has begun to rise in the new countries, already we see the American farmer is in his turn being compelled to resort to fertilizers; and with each rise in prices the intensive farmer can recoup himself for an increased outlay. The future, too, lies with intensive farming; every year the ratio of the cultivable land to the population of the world shrinks; every year science puts fresh resources in the hands of the farmer. In the United Kingdom for some time the stream may still run backwards and the more expensive forms of arable cultivation continue to be replaced by grass which demands no outlay, because as long as ours is the one market open to the competition of all other countries selling agricultural produce, prices are still liable to such wreckage as frightens the home grower out of the business; still, in the end, whatever agriculture survives in this country will be forced into more and more intensive methods by the increased scarcity of the land. As it is, the specialist farmers in Great Britain—the potato growers, the market gardeners, the hop growers—have reached a pitch of cultivation which is hardly to be paralleled elsewhere.

Intensive farming implies the use of fertilizers; still more it implies, or should imply, skill and knowledge in using them.

If this book is to have any justification for its existence, it will be by helping men to a greater skill and knowledge in the use of their fertilizers and manure. There is no lack of books which give an account of the origin and composition of fertilizers: my object is rather to make the reader understand their mode of action and their relation to particular crops and soils. For it is only by understanding the why and the how that a farmer can properly adjust his manures to his soil and his style of farming: he must to some extent reason the scheme out for himself, he cannot simply be told.

The scientific man is always being asked to arrange his experiments to demonstrate the *best* way of growing this or that crop, by the best being implied the cheapest: farming visitors to Rothamsted are often inclined to suggest that the plots, if interesting, are not "practical." After sixty years of work they rather expect to see the absolutely cheapest form of manuring each crop set out once and for all. But in practical farming there is no "best" way of doing things; the mere fact that the weather of the coming season is unknown makes it impossible to specify the absolutely right course either in cultivation or in manuring. The question even of the best manure for a given crop is complicated by the manner in which every farm differs somewhat from every other, not merely in its soil and climate, for these matter less than is commonly supposed, but in its object and management. One man aims at crops, another man gets his money back by his stock; one man has only to pay 15s. an acre rent, another has to get twice as much out of his land before he touches a profit; one man's markets are such that he can repay himself for an outlay of £3 an acre for fertilizers on his root area, whereas another man could not afford 20s.; no one receipt can be handed out to suit all these different men.

The object, then, of the scientific man should be to lay down principles which the practical man in his turn must learn to apply to his own conditions; success is only possible when he, too, does some thinking. Furthermore, the object of experiments should be to provide knowledge that can be thus applied to other conditions, and an experiment is practical just in so far as it carries out its avowed object, which is to lead men into a sound and fruitful way of thinking on the question at issue.

It is in this respect—the elucidation of general principles—that the Rothamsted experiments have proved so exceedingly valuable; though initially laid out to test certain definite questions about the nutrition of crops, the answers to which have long since been absorbed into farming practice, the design was so sound and the continuity of the record has been so rigorously maintained that the results now afford an instructive commentary on the whole range of the science of crop production. We have by no means come to the end of the lessons the Rothamsted experiments can teach: every new theory, each extension of our knowledge, finds an unsuspected criticism or an illustration in the records that are still accumulating.

I have, in consequence, throughout this book used very freely the results of the Rothamsted experiments; and if the conclusions I have drawn do not always square with popular opinion, I have none the less set them out in the hope that other experimenters would thereby be led to check or revise them. Agricultural chemistry is still cumbered with a good many *a priori* deductions resting upon a very slender foundation—first approximations to the truth which fail because they do not take all the factors into account; it is about many of these opinions that the Rothamsted results suggest scepticism.

The book is intended for farmers and for the senior students and teachers in our agricultural schools. I have therefore kept the language as non-technical as possible, though some elementary knowledge of chemistry has to be assumed. If sometimes, as in Chapter XI, I may seem to have gone rather far in the discussion of theoretical questions, it is in pursuance of my main idea that it is only by thinking about the rationale of manuring we can arrive at right practice. And as the book is intended for those who are using or going to *use* fertilizers, I have not troubled to say much about their manufacture, nor have I dealt at all with their analysis: these are both technical matters outside the business of the farmer.

I have meant this to be a companion to my book, *The Soil*; they are both written for the same audience, and on similar lines. I hope later to complete the series, by a third book, dealing with the chemistry of the growing plant.

A good deal of the material in the book has already been utilized and, in part, published in a course of Cantor Lectures delivered before the Society of Arts in 1906, and again in a course of lectures delivered at Cornell University to the Graduate School of Agriculture of the United States Department of Agriculture in July, 1908. In this way, much of the substance of Chapters II, III, IV, V, and VI, has already been printed in the *Journal of the Society of Arts*, the greater part of Chapter VII has appeared in the *Journal of the Board of Agriculture*, and Chapter XI, the last of my American lectures, was printed in *Science*.

I have drawn so freely upon my friends for information, that it seems invidious to single out for thanks some more than others: but I owe much to Dr. E. J. Russell of this Laboratory, who has read and criticized parts of the text; to Dr. J. A. Voelcker, who has so often placed the results of his wide experience at my disposal; to Mr. H. Voss of the Anglo-Continental Guano Co., and to Mr. T. Elborough, of the Lawes Chemical Manure Co., who have furnished me with many facts and figures respecting the trade in fertilizers; and once again, to Mr. G. T. Dunkley, of the Rothamsted Laboratory, who has been indefatigable in verifying references and in securing the accuracy of the many figures and tables the book contains.

A. D. HALL.

CONTENTS

PAGE

CHAPTER XI

FERTILIZER REQUIREMENT AND THE TRACE ELEMENTS

CHAPTER XII

SYSTEMS OF MANURING CROPS

CHAPTER XIII

THE VALUATION AND PURCHASE OF FERTILIZERS

CHAPTER XIV

THE DISTRIBUTION OF FERTILIZERS AND FIELD EXPERIMENTATION

LIST OF ILLUSTRATIONS

Hall, Sir (Alfred) Daniel (1864-1942), educationist, administrator, and scientific research worker, was born at Rochdale 22 June 1864. He was educated at Manchester Grammer School, of which he was a scholar, and at Balliol College, Oxford, where he was elected to a Braekenbury scholarship in natural science at the age of seventeen. There he stroked the college eight, debated in the Union, and took a first class in natural science (chemistry) in 1884.

Going down from Oxford, Hall look up teaching, he was senior science master at King Edward's School, Birmingham, in 1888, amongst his initiates being Ernest William Barnes, later bishop of Birmingham. In 1891, he became lecturer under the Oxford university extension delegacy in Sussex, Surrey, and Kent. Here began the principal work of his life, when he introduced the teaching of agricultural chemistry into his courses of lectures on fundamental science to country classes. In 1894, he was installed as first principal of the South Eastern Agricultural College at Wye, Kent. He collected a small staff of well-qualified and enthusiastic university graduates, and together they evolved a system of agricultural education upon which have been based all the departments of agriculture of the universities and the agricultural colleges founded since that time. Simultaneously, he attacked the problem of agricultural research. There were then no public advisory services available to farmers and by the organization of winter lectures to farmers and of experimental work on their farms the educational work of the College was extended.

In 1902 Hall, the obvious selection, succeeded to the directorship of the agricultural experimental station at Rothamsted, in Hertfordshire. During the previous half-century the place had become world famous by the work of its founder Sir J.B. Lawes and his friend and colleague Sir J.H. Gilbert (qq.v.), and the time had come for a complete review of its organization, its scope, and its achievements. This was the task to which Hall applied himself, and in the following years he supplied the agricultural student and the more educated farmers with a series of publications illuminating the fundamental problems of soil science and food production. The first was his classic on *The Soil* (1903), and this was followed by *The Book of the Rothamsted Experiments* (1905) summarizing the life-work of Lawes and Gilbert. *Fertilizers and Manures*

(1909) and *The Feeding of Crops and Stock* (1911) were students' textbooks and in the latter year was published also *The Agriculture and Soils of Kent, Surrey and Sussex*, embodying the results of a long-term investigation and survey undertaken in his Wye College days in collaboration with (Sir) E. John Russell. There followed *A Pilgrimage of British Farming* (1913), the record of surveys made by Hall in three previous summers in the company of T.B. Wood, professor of agriculture at Cambridge, and E.S. Beaven, barley expert to Arthur Guinness & Co.

In 1910 Lloyd George invited Hall to become a member of the Development Commission. This appointment was part-time and unpaid, but two years later, the Rothamsted Experimental Station being by that time satisfactorily reorganized, he resigned from it to devote himself wholly to the work of the Commission. This consisted, for the most part, in supplementing the work of the Board of Agriculture and Fisheries in organizing agricultural education and research, particularly by grants in aid and establishing centres for specialist research, and by advising farmers through the county councils. Other activities falling within the Commission's scope were the promotion of agricultural co-operation; the reclamation of sub-marginal land; and the examination of the economics of new crops, such as sugar-beet and tobacco.

The outbreak of war in 1914 found Hall in Australia, as president of the agriculture section of the British Association. Hurrying back to England, he was soon involved in the problem of feeding the nation at war. He was a member of the committee, over which Lord Milner (q.v.) presided, to advise upon measures for the increase of production. This recommended guaranteed prices for wheat, but these were not introduced until 1917. In 1916 R.E. Prothero (later Lord Ernle, q.v.) was appointed president of the Board of Agriculture, and at his request Hall left the Development Commission in 1917, to become secretary of the Board. In this capacity he was largely concerned with the president's war-time measures for stimulating food production in the face of the submarine menace, and for the post-war efforts of the Government to carry through a policy for stabilizing the position of the agricultural industry. Considerable success attended the former, but the latter broke down with the return of peace and plenty.

From 1920 to 1927 Hall continued at the Ministry of Agriculture (as it had now become) as chief scientific adviser. His field of work was a wide one, extending from the completion of the chain of research stations covering all branches of agricultural science to the respiration of the Government at international gatherings.

Hall had long been a member of the governing body of the John Innes Horticultural Institution, at Merton, and in 1927 he succeeded William Bateson (q.v.) as director. He remained at Merton for twelve years, retiring in 1939 just before the outbreak of war, and the position which he established for the Institution was another proof of his versatility as a scientist and administrator. He was not a geneticist by training, but his capacity to appreciate the nature of the work enabled him to select and encourage a team of young botanists who made the John Innes Institution famous throughout the world. During this time Hall produced two notable books, one, on *The Apple* (1933) in collaboration with his colleague, Mr. M.B. Crane; the other, a monograph on *The Genus Tulipa* (1940). Other books written by him were *Agriculture after the War* (1916), *Digressions of a Man of Science* (1932), and a book for children, *Our Daily Bread* (1938). Shortly before his death, he produced *Reconstruction and the Land* (1941) in which he formulated a national agricultural policy for the future, based on a lifetime's experience of agricultural practice, administration, and research.

Many public honors and distinctions were bestowed upon him. He was elected a fellow of the Royal Society in 1909, but refused an offer on nomination for its presidency some years later. He was appointed K.C.B. in 1918, and honorary doctorates were conferred upon him by the universities of Oxford, Cambridge, and Aberdeen. He was Rede lecturer at Cambridge and Health Clark lecturer in London in 1935. He was chairman of the Kenya agricultural commission in 1929, and a member of the Agricultural Research Council and of the Economic Advisory Council. In 1939 he was elected to an honorary fellowship at Balliol, and in the same year, on the occasion of his seventy-fifth birthday, a group of his friends presented him with a book of essays entitled *Agriculture in the Twentieth Century.*

Hall was a man of extraordinary versatility, who touched life at many points. Music, painting, and literature were constant pleasures to him, and he was a collector of early Chinese pottery, upon which he was a recognized authority. A discriminating palate for French wines and a devotion to fly-fishing rounded off a personality unusually complete and well balanced.

CHAPTER I

INTRODUCTORY

Historical—The nutrient solution—The nutrients in the soil.

Historical

THE word "manure," when first met with in English, possessed a much wider significance than it does to-day. Of the same origin as manœuvre, it meant, primarily, to work by hand, and it is used in that sense by Defoe in *Robinson Crusoe*—"The land which I had manured or dug"; but it also took on the extended meaning of any process or material by which the land could be ameliorated. In the seventeenth and early eighteenth centuries this latter sense alone began to prevail; agricultural writers enumerated such substances as chalk, lime, marl, burnt clay, as manures, and began to speak of the operations of cultivation as tillages or husbandry; and more recently the tendency has been to restrict the employment of the term even further, confining it to the natural substances possessing a direct fertilizing value. Farmyard manure is the typical "manure"; marl and chalk are no longer regarded as manures but rather as soil improvers; while substances like basic slag or sodium nitrate, which simply supply one or other element in the nutrition of a plant, are termed "fertilizers" rather than artificial manures. The distinction is not, however, very clearly drawn, and manure and fertilizer are generally and unconsciously used as interchangeable terms.

It is impossible to assign a period to the discovery of the fertilizing properties of the excrement of animals: agriculture must be almost coeval with the human race; and that tissue of experience and observation which reaches us as the tradition of farming—the stock-in-trade of the practical man—began to form long before letters existed by which it could be recorded. At any rate, when in Roman times we began to get some record of agricultural practices, we find that not only was the value of dung recognized, but that the virtues of certain other materials like marl, had been established. Even the fertilizing effect of a crop of vetches or lupins upon the succeeding wheat crop was sufficiently well known to be related, not only by professed agricultural writers like Varro and Columella, but also by a poet like Virgil. But to whatever point the knowledge of manures had reached in the time of the Romans, for a long time it made no further advances and bade fair to be utterly lost with the irruption of the barbarians. When the new peoples emerge again in Europe, after the great movements of the races, we mostly find them practising the Germanic common field system of agriculture, with its rotation of wheat, beans or barley, and fallow, followed up by general grazing over the whole area—a system which lends no encouragement to the use of substances like manures for the improvement of the land.

Doubtless the old traditions did not perish in the Romance countries, but as before were handed down from one generation to another; as long as corn and wine continued to be cultivated the immemorial precepts concerning their management would linger about the countryside and be treasured in the memories of the workers in the fields. But during the Dark Ages this kind of knowledge sank below the level of whatever literature was being written; it had to diffuse slowly from the remains of Roman civilization among the invading peoples, and it is only by chance that we get any record of what the countryman did or thought. In many English tenures we find that the flocks of the tenants had to be folded on the lord's land at night, the manure thus brought being one of his most valued privileges; while in Walter de Henley's *Husbandrie*, the great mediæval treatize on the duties of a land agent, we find instructions for the preservation of dung by the use of litter and marl. The manure thus obtained was to be stored in a heap and preferably applied to sandy land. From mediæval times also we derive such maxims as the Flemish proverb:

> **Point de fourrage, point des betail,**
> **Point des betail, point de fumier,**
> **Point de fumier, point de fourrage.**

When, with the general resurrection of learning at the Renaissance, we once more get books on agriculture, we find that either old tradition or the experience of men of an inquiring turn of mind, who had been trying all sorts of things on their land, had already built up a certain knowledge of manures and manuring. The value of marl and chalk, of woollen rags and ashes, was certainly known in the sixteenth century; men had even begun to reason a little on the mode of action of manures. For example, Bernard Palissy the potter, in his *Recepte Véritable*, published in 1563, not only recommends the use of marl and lime, but can assign a reason for the value of ashes, and shows that the richness of farmyard manure resides in the portion soluble in water: "Et ainsi la paille estant bruslee dedans le champ, elle seruira d'autant de fumier, parce que elle laissera la mesme substance qu'ell auoit attiree de la terre . . . ;" and again, "au lieu où ledit pilot de fumier aura reposé quelque temps, ils n'y laisseront rien dudit fumier, avis le jetteront deçà et delà, mais au lieu ou il aura reposé quelque temps, tu verras qu'apres que la blé qui aura esté semé sera grand, il sera in cest endroit plus espes, plus haut, plus verd et plus droit. Par là tu peux aisement cognoistre que ce n'est pas le fumier qui a causé cela, car le labourer le jette autre part; mais c'est que quand ledit fumier estoit au champ par pilots, les pluyes qui sont suruenues, ont passé à travers des dits pilots, et sont descendu à travers du fumier jusqu'à la terre, et en passant, ont dissout et emporté certains parties du sol qui estoit audit fumier."

If, then, by the sixteenth century we find written evidence of the knowledge of the fertilizing properties not only of dung but of other waste substances, we may safely push back the original discovery of the properties of these bodies to a much more remote epoch, if such a term as discovery can properly be applied. Just as happens to-day, this or that man tried an experiment or noticed the result of an accident which caused him to report well of the action of some substance on his crops; his opinion would often be mistaken and often, again, it would be forgotten; but occasionally it would be repeated and find confirmation, until it acquired the wide circulation and staying power of a farming tradition and passed more or less into the common routine. Even at the present time there are many beliefs and practices more or less current among farmers, which science has neither verified nor disproved, and which may be either examples of sound observation or only imperfect generalizations. Such opinions require to be examined with the utmost care and openmindedness, for even when correct they are of no final use to agriculture until they have been explained and absorbed into the general stream of scientific knowledge. The value of many fertilizers must have been observed and lost sight of over and over again, because of the lack of any general theory to serve as a touchstone and discriminate between the true and the false reports. So, despite the experience that was accumulating respecting the fertilizing value of this or that substance, no real progress towards a theory of manuring was made until the close of the eighteenth and the beginning of the nineteenth century.

Before the development of a science of chemistry it was naturally impossible to form any idea of how a plant came to grow; while the nature of the plant itself, and of the air, water, and earth were unknown, no correct opinion could be reached as to how the latter gave rise to the former. In spite of Palissy's very sound conclusions as to the salts plants draw from the ground, Van Helmont described an experiment to show that a tree is made out of water alone. Jethro Tull, arguing from his hoeing experiments, concluded that manures were unnecessary, for the soil, if only stirred up enough and exposed to the air, will provide all that the plant requires. Even so late as 1810 we find Thaer writing that there is no doubt that the fallow absorbs or attracts the fertilizing properties of the atmosphere.

The true theory of the nutrition of the plant begins very soon after the discovery of the composition of the air. "Thus, Priestley observed that plants possessed the faculty of purifying air vitiated by combustion or by the respiration of animals; and he having discovered oxygen, it was found that the bubbles which Bonnet had shown to be emitted from the surface of leaves immersed in water consisted chiefly of that gas. Ingenhousz demonstrated that the action of light was essential to the development of these phenomena, and Sennebier proved that the oxygen evolved resulted from the decomposition of the carbonic acid taken up."

Following up these results, de Saussure demonstrated with as much quantitative accuracy as was then possible that the oxygen which was split off by the leaf was contained in the carbonic acid, and that the gain in weight of the plant was practically represented by its carbon, combined with the elements of water to make up such carbohydrates as sugar and starch. De Saussure further arrived at very clear ideas as to the source and value of the ash constituents of plants: the nitrogen, which he also pointed out was an invariable constituent of plants, he considered to be derived either from the ammonia in the atmosphere or from the organic matters in the soil.

Sir Humphry Davy, in his lectures before the Board of Agriculture from 1802 to 1813, practically adopted de Saussure's views, and emphasized the importance of the ash constituents, which could come neither from the air nor water, as he yet thought it necessary to demonstrate. Though Davy made no advances towards ascertaining the relative importance of these substances, and was by no means certain that the plant derived all its carbon from the atmosphere, his lectures did much to pave the way for the adoption of a sound theory.

Thaer, about the same period, was still attributing the chief share in the nutrition of the plant to the humus in the soil, which he showed to contain hydrogen, nitrogen, sulphur, and phosphorus. Though knowledge of the composition of plants, soils, and manures continued to accumulate, as seen in the work of Sprengel and Schubler, the next step forward was due to Boussingault, who was the first man to undertake field experiments on a practical scale. Farming his own land at Bechelbronne, Alsace, from 1834 onwards, he systematically weighed crops and manure and analysed both so as to obtain a balance-sheet showing the quantities of carbon and nitrogen added in manure and removed in the crops. In 1838 he demonstrated on a working scale the enormous amounts of carbon which are assimilated by the plant from the atmosphere —far greater quantities than the humus in the soil could continue to supply. Boussingault's experiments led him to conclude that the plant derives its nitrogen from the soil, though he also showed that in certain rotations more nitrogen is removed in the crop than is supplied in the manure.

It is to Liebig, however, that we must attribute the chief impulse which agricultural chemistry has received; though he made little original contribution himself to the theory, adopting in the main the conclusions that arose from the work of Priestley, Ingenhousz, Sennebier, and de Saussure, he was the man who drove home to the minds of scientific men and of farmers, the true theory of plant nutrition. In his report to the British Association, published in 1840, and his *Chemical Letters*, he laid down the general principle that the carbon compounds, which constitute about 95 per cent. of the dry matter of the plant, are derived from the atmosphere, and that if the plant be supplied with the 2 per cent. or so of mineral constituents which are found in its ash, it will then draw upon the atmosphere for all the other materials the crop ultimately contains.

Coming at a time when much intellectual interest was directed towards agriculture and backed by his great scientific reputation and his commanding personality, Liebig's views aroused instant and general attention; they became the foundation of both practical experiment and scientific research, and were the starting-point of a prolonged controversy. In several respects Liebig's views required modification; he seemed to consider all the ash constituents were of equal importance, and even that one element like sodium could replace any similar one like potassium ; he regarded the composition of the plant's ash as determining its need for fertilizer —a point which will be considered later; and he misapprehended the part played by nitrogen compounds in manuring. At that time, owing to imperfections in the methods of analysis, very exaggerated ideas were current as to the amount of ammonia naturally present in the atmosphere; the rain was believed to bring down 30 or 40 lb. per acre per annum of combined nitrogen instead of the 3 or 4 lb. which we now know to be contributed to the soil, and to this source Liebig was disposed to look for the nitrogen in the plant. He stated that manures containing nitrogen certainly stimulated growth, because they fermented and increased the proportion of ammonia in the air round the plant; but in the main nitrogenous manures were unnecessary, for full crops could be grown if only the constituents removed in the ash were annually returned to the soil.

In particular, these views on the nutrition of plants and the part played by the nitrogenous manures, did not commend themselves to John Bennet Lawes, a young Hertfordshire landlord who had recently come into possession of the family estate of Rothamsted, near St. Albans, and had already begun to try manurial experiments on a small scale. In 1843 the experiments took more systematic form and Lawes obtained the services of Joseph Henry Gilbert, a chemist who had worked under Liebig at Giessen, to conduct them, thus inaugurating the field trials which have continued without a break to the present day. The immediate result of the Rothamsted experiments was to demonstrate the necessity of a supply of combined nitrogen, the yield being in fact roughly proportional to the amount of combined nitrogen added as manure. If only the mineral constituents of the ash were supplied, the crop fell away rapidly as soon as the reserves of active nitrogen in the soil had become exhausted. Lawes, Gilbert, and Pugh further repeated, with great exactitude, a series of laboratory experiments initiated by Boussingault, and demonstrated that the ordinary plants of the farm were incapable of utilizing the free nitrogen of the atmosphere, but only took up nitrogen in a combined form from the soil or the manure. Lawes and Gilbert were fiercely attacked by Liebig; but as far as his views can be extracted from his writings, there was no very great difference between the opinions held by the rival parties. Liebig laid the chief stress on the need for the mineral manures, whereas the other investigators were more concerned to demonstrate the importance of nitrogen. Liebig also seems to have thought that the leafy crops, like clover or roots—the so-called restorative crops—could gather nitrogen from the atmosphere and dispense with any supply in manure. But in addition to establishing the value of nitrogenous manures, the Rothamsted experiments also settled in a practical fashion the question of which of the ash constituents were indispensable to the plant and were necessary constituents of a complete manure. The fundamental necessity of phosphorus and potassium, and the non-essential nature of sodium, magnesium, and silicon as manure constituents were soon established, and the experiments began at once to bear fruit in the way the various fertilizers, then being discovered and put on the market, were utilized by farmers. At the same time the methods of plant culture in weak solutions of known salts were evolved, and in the hands of Boussingault, Knop, Stohmann, and others, demonstrated with all the precision then possible that, of the elements found in the plant, only compounds of nitrogen, phosphorus and sulphur among acids, and potassium and calcium with a trace of iron among bases, were absolutely essential to the nutrition of the plant.

There was, however, still one point which remained somewhat inexplicable—the gain in combined nitrogen which seemed to take place when certain crops of the leguminous order were grown. Cases were recorded where more nitrogen was found in the crop than was supplied in the manure, and yet the soil on which the crop had been grown itself showed an increase of nitrogen.

Boussingault, in his earliest investigations, had shown that in certain rotations which included clover or lucerne, more nitrogen was removed in the crop than had been supplied in the manure, and many of the Rothamsted results could only be explained on the assumption that the roots of such crops ranged exceptionally deep and drew upon stores of subsoil nitrogen unavailable for other plants, thus leaving the upper soil the richer for their growth, since the roots and stubble, in which this subsoil nitrogen has been accumulated, decay near the surface. It was not until 1886 that these difficulties were cleared up with the discovery by Hellriegel and Wilfarth that leguminous plants do "fix" the atmospheric nitrogen by the help of certain bacteria living in symbiosis with the root of the leguminous plant. The leguminous plant, however, will also feed upon combined nitrogen in the soil like any other plant, and the failure of Lawes and Gilbert to detect any nitrogen fixation in their laboratory experiments with beans and clover, was due to the great care taken to shut out any intrusion of foreign matter during the experiments, thus preventing the leguminous plants from becoming inoculated with the bacteria causing fixation. In a measure, the discovery of Hellriegel and Wilfarth, which formed the starting-point of much further research, may be taken to have justified some of Liebig's arguments, although the mechanism by which the nitrogen fixation is brought about—by bacteria living in concert with the higher plant—would have been entirely foreign to his way of looking at things, just as it was to Lawes and Gilbert, who thus unhappily missed the clue which would have explained many of their results.

It has already been indicated how impossible it is to recover the date of the original discovery of the fertilizing value of the substances we now call manures; only by an occasional allusion in the older books can we find that particular materials were in common use at the period of the writer. Blithe's *English Improver*, published in 1653, mentions the value of rags, wool, marrow bones or fish bones, horn shavings, soot, and wood ashes; and Evelyn, writing a few years later, adds also blood, hair, feathers, hoofs, skin, fish, malt dust, and meal of decayed corn, so that a knowledge of the value of these materials must have been widespread.

Early in the nineteenth century we begin to hear of guano from Peru, though the first importation did not take place until 1840. The importation of sodium nitrate from Chile had begun a year or two before; its value as manure was for a time in doubt, though as early as 1669 Sir Kenelm Digby had recounted an experiment to show how much barley plants were benefited by watering with a weak solution of nitre, and Evelyn in 1675 had written: "I firmly believe that were saltpetre to be obtained in plenty, we should need but few other composts to meliorate our ground."

The employment of ammoniacal salts seems to have begun entirely upon theoretical grounds; de Saussure had attributed the nitrogen of vegetation to the ammonia in the atmosphere, and in this he was followed by Liebig; fortunately, about the same time, the manufacture of coal-gas gave to the world a cheap source of ammonium salts. Lawes had already been trying them before Liebig's paper of 1840, and when the Rothamsted experiments were definitely started in 1843, a mixture of ammonium chloride and sulphate became their standard nitrogenous manure.

The use of mineral phosphates as manure begins with Lawes' superphosphate patents in 1842, although no mineral phosphates were available on a large scale until Henslow's discovery of the coprolite beds of Cambridgeshire in 1845. Putting aside the various methods adopted for the utilization of waste products like slaughter-house refuse, no further novel manurial substances can be said to have been introduced until the development of the Stassfurt deposits of potassium salts, which began about 1860, and the discovery of basic slag in 1879. Various processes for bringing atmospheric nitrogen into a combined form have been the outstanding advance of the twentieth century.

The Nutrient Solution

Since the object of employing manures is to improve the nutrition of the plant, it is desirable to consider at the outset how the plant is constituted. Water makes up by far the greatest portion of the living plant—some 80 or 90 per cent. of the total weight—while carbon accounts for about 50 per cent. of the dry matter, in the form of various carbon compounds. A large proportion of these compounds consists of carbohydrates, such as sugars, starch and cellulose, which, in addition to carbon, contain hydrogen and oxygen in the same proportions as are found in water. The framework or skeleton of the plant is cellulose. The fats are also built up from the same three elements, but there is a very important class of substances, commonly grouped under the term crude protein, containing nitrogen and usually phosphorus and sulphur as well as carbon, hydrogen and oxygen. The actual amount of nitrogen in the dry matter of plants varies considerably according to stage of growth, but is commonly between 1 and 3 per cent. When the plant is burnt, the ash is found to amount to 5 per cent. or more of the dry matter and contains potassium, sodium, calcium, magnesium and iron among the metals and chlorine, sulphur, phosphorus and silicon among the non-metals. Traces of many other elements are also found and, although they were all regarded at one time as quite unessential, it is now recognized that many of them are required for healthy plant growth. The question of the trace elements will, however, be dealt with more fully in Chapter XI.

Carbon, then, is the chief element in the economy of the plant and the carbon dioxide in the atmosphere and water from the soil are the raw materials in carbon assimilation. The amount of carbon dioxide in fresh air is only about 0·03 per cent. by volume; hence enormous quantities of air are required to supply the needs of plants for carbon. But carbon dioxide is continuously being formed by the combustion of carbon compounds, and by their decomposition by various organisms so that its concentration in the atmosphere remains fairly constant. Considerable amounts are produced, for example, in the soil, which helps to maintain the concentration in the neighbourhood of the leaves of plants. Nevertheless, carbon dioxide is sometimes a limiting factor in the growth of plants and, under favourable conditions, crop yields have been considerably raised by increasing its concentration artificially.

The agent in the assimilation of carbon is the green pigment chlorophyll of the chloroplasts. These small bodies are found mainly in the elongated cells of the palisade tissue lying below the epidermis. The carbon dioxide is absorbed through the open pores of the stomata in the epidermis, and the amount absorbed is roughly proportional to the number of stomatal openings. In the case of a sunflower leaf, the rate of diffusion in bright sunlight is about 10 to 15 c.c. per 100 sq. cm. of leaf surface per hour. But the assimilation is governed by plant species, illumination, temperature and the other factors controlling plant growth. The carbon dioxide is dissolved in the water on the cell walls before reaching the chloroplasts, where it undergoes a chemical change under the action of light. Since radiant energy is absorbed by the chlorophyll and converted into chemical energy, the reaction is called a photosynthesis. It can be effected by light of all wave lengths but the yellow to red end of the spectrum is the most efficient. About 70 per cent. of the sunlight falling on the leaf is absorbed, the remainder being reflected or transmitted, but only a small fraction (about 2 or 3 per cent.) of the light absorbed is used in photosynthesis. When the illumination is low, photosynthesis is proportional to the light falling on the leaf, but if illumination and carbon dioxide are plentiful then temperature plays a most important part on account of its effect on other reactions involved in assimilation.

The complete mechanism of the process of photosynthesis is still uncertain, but the first identifiable products of the

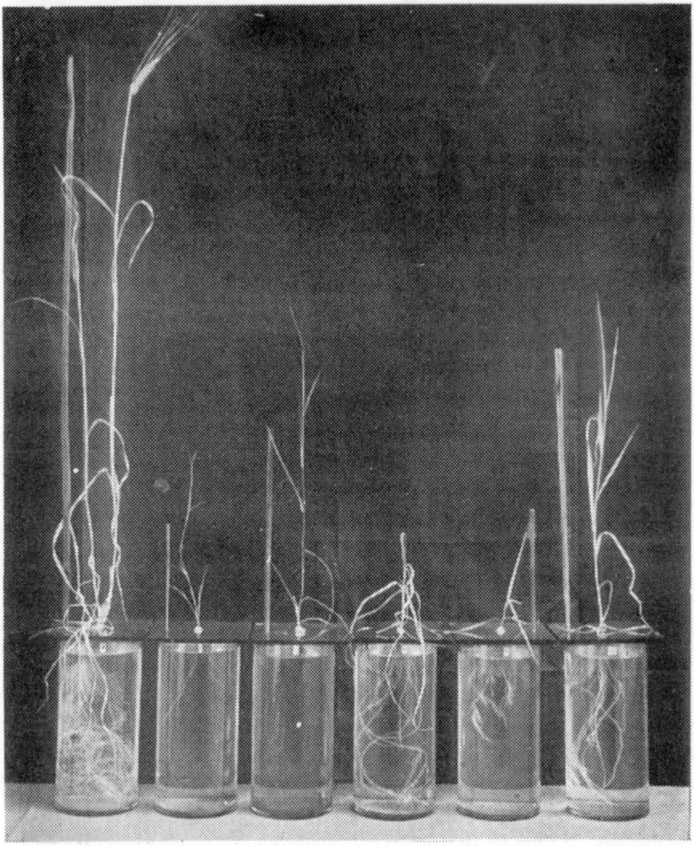

1 2 3 4 5 6

FIG. 1.—WATER CULTURES OF BARLEY

1. Complete Manure
2. No Nitrogen
3. No Phosphoric Acid
4. No Potash
5. No Lime
6. No Magnesia

process are a phospho-glyceride and simple sugars, while the volume of oxygen set free into the atmosphere is equal to the volume of carbon dioxide absorbed.

Although the compounds of carbon make up so much of the dry matter of the plant, the comparatively small quantities of the other elements are all important to the growth processes. The part which they play can easily be illustrated by means of water cultures as shown in Fig. 1. The roots of young seedlings dip into a dilute solution of salts of the essential elements. Many different complete nutrient solutions are used by investigators, and the following is a typical example:

	grams per litre.
Potassium Nitrate	1
Potassium Dihydrogen Phosphate	0·5
Calcium Sulphate	0·5
Magnesium Sulphate	0·5
Sodium Chloride	0·1
Ferric Citrate	0·02

For demonstration purposes, the slight impurities in the water and reagents take care of the trace elements, but in precise work specially purified substances are employed and definite small amounts of compounds containing boron, manganese, zinc and copper are added—less than 1 part per million of solution is required in each case. Sometimes the solution is allowed to percolate slowly through a pot containing pure quartz sand in which the plant is anchored; sometimes a much more dilute solution is employed and is renewed at short intervals to maintain a suitable concentration of elements for the growing plant. But whatever the experimental details, provided the nutrient solution contains all the elements normally required and precautions are observed with respect to light, temperature and aeration of the roots, the plant will complete its life cycle by assimilating carbon dioxide, producing dry matter, setting flowers and ripening healthy seed. Its growth is normal so that the method provides a means of following the course of nutrition free from the complex processes which go on in the soil.

The composition of the nutrient solution may be altered by omitting one element at a time, and Fig. 1 illustrates such a trial. It shows how, when any one of the major essential elements is absent, the plant is unable to grow after it has used up the supply of that particular element in the seed. The plant must be able to obtain through its root system all the constituents indispensable to growth, and nothing can compensate for the lack of one of them. Some regular constituents of plant ash such as sodium and silicon are apparently not essential, but nitrogen, phosphorus and sulphur, calcium, magnesium, potassium and iron definitely are essential, together with certain other elements in minute amounts. Hence the soil must be able to provide these elements. They are carried into the plant in the transpiration current of water which enters the root hairs and is eventually evaporated from the leaves and other growing surfaces. Under normal conditions a plant transpires from 200 to 500 pounds of water for every pound of dry matter it produces; the lower figure prevails in a humid climate like our own, whereas the higher one is found in drier countries. One inch of rain over an acre amounts to 113 tons, and so the production of 2 tons of dry matter involves something like 5 or 6 inches of rain. The concentration of the soil solution is very variable but generally small, and the various salts present may be regarded as ionized, that is to say each is dissociated into a negative and a positive ion. Any dilute solution, therefore, contains a balance of negatively charged ions or anions, such as nitrate or phosphate, and positively charged ions or cations, such as ammonium or potassium. Hence nitrogen can enter the plant either as a nitrate anion or an ammonium cation, phosphorus enters the plant as a phosphate anion and the metals enter as the metallic cations calcium, magnesium and so on.

The plant does not necessarily absorb a salt as a whole, but absorbs the individual ions according to equilibria set up between the external and internal conditions. For example, one ion may be immediately translocated on absorption to take part in some chemical action in the plant and so does not get a chance to accumulate and retard the rate of entry. Another ion may not be required in the metabolism so that its absorption is slowed down. In this respect roots may be said to show a selective action. Water or ions may pass independently through the membranes of the root hairs and there may actually be a movement against the concentration gradient; in other words, there may be a much higher concentration of an ion within the plant than outside it. Plants can absorb, for example, sufficient amounts of phosphate from a solution containing only a few parts of phosphate per million of solution. In fact, the solution in the plant may be quite different from that outside in both concentration and composition. This is because different ions have different degrees of mobility to effect an entry, and are subject to different forces in their removal from root to shoot. Usually the cation is absorbed more rapidly than the anion, which may lead to the external solution becoming acid; but sometimes the reverse takes place as, for example, with nitrates, and the solution may become alkaline for nitrate is very rapidly absorbed and utilized by the plant and leaves excess cation behind which forms an alkali.

The balance of ions in the external solution is always maintained by equivalent replacement of ions from the root cells, and this ionic exchange across the root hair membrane is an important factor in the mechanism of absorption. The whole process, however, is influenced by interionic relationships, one ion being able to retard or accelerate the absorption of another. The proportions of ions in the external solution may be altered through a relatively wide range without appreciably affecting the rate of growth or yield of the plant, but the chemical composition of the plant may be markedly altered by quite small changes in the ionic balance. It is the effectiveness of this physiological balance of ions which determines a well-regulated growth and healthy development. Even an otherwise adequate solution may be deficient in one element at a particular stage of growth, or excess of one ion may render another unavailable at a critical point in the development of the plant.

The Nutrients in the Soil

No single chemical criterion can assess the supplying power of soils for a specific nutrient. The soil is not to be regarded merely as an inert medium to anchor the plant and convey nutrients to it when convenient, but is itself an enormous potential reserve of plant food. For example, Table 1 shows the composition of a soil about which so much knowledge has been accumulated.

TABLE 1.—ANALYSIS OF THE SOIL OF BROADBALK FIELD, ROTHAMSTED. UNMANURED FOR 50 YEARS.

	Per cent.		Lb. per Acre.
Loss on ignition	4·20	...	...
Containing Carbon . . .	...	0·89	22,250
,, Nitrogen . .	...	0·10	2,500
Matter soluble in Hydrochloric Acid	12·53	...	...
Containing Soda . . .	...	0·06	1,500
,, Potash . . .	...	0·27	6,750
,, Magnesia . .	...	0·36	9,000
,, Lime . . .	...	2·49	62,250
,, Alumina . . .	...	4·49	112,250
,, Oxide of Iron . .	...	3·40	85,000
,, Phosphoric Acid .	...	0·11	2,750
,, Sulphuric Acid .	...	0·05	1,250
,, Carbonic Acid. .	...	1·30	32,500
Undissolved Siliceous Matter .	83·27	...	...

As usual, the greater part of the soil consists of insoluble siliceous matter, of which no account need be taken; there is, further, a certain amount of organic material, important as containing a store of nitrogen which may eventually reach the plant. In addition there are various substances which are dissolved by the acids used in the analytical process, and these include precisely the substances that have already been indicated as constituents of the ash of plants. Amongst the metals are calcium, magnesium, potassium, sodium, with iron and aluminium in very much larger amounts; sulphur and phosphorus, chlorine and silicon are non-metals. Read as percentages, some of these amounts seem small enough but they represent enormous quantities of material in the soil, for the layer of soil at Rothamsted, nine inches deep, which was sampled for analysis, weighs, over the area of one acre, rather more than two and a half million pounds. Translating, then, the percentages into pounds per acre, 0·1 per cent. of nitrogen becomes 2,500 lb., 0·11 of phosphoric acid becomes 2,750 lb., and the potash rises to 6,750 lb.; these quantities are in the surface soil only, and distinct from those of the lower layers into which the plant roots penetrate freely. A comparison of the materials in the soil with those taken away by ordinary crops at once leads to results which seem paradoxical; the stock of plant food in the soil is so much greater than any requirements of the crop that further additions of the same things in the shape of fertilizers would seem to be needless. The accompanying table (2) shows the amounts of various materials per acre which are, on the average, drawn from the soil by various crops at Rothamsted.

Roughly speaking, an average soil contains enough plant food for a hundred full crops, yet without fresh additions of

TABLE 2.—SOIL CONSTITUENTS CONTAINED IN AVERAGE CROPS.

	Wheat.	Barley.	Swedes.	Mangolds.	Hay.
Crop . .	Tons. 2·2	Tons. 2·0	Tons. 16·1	Tons. 30·1	Tons. 1·5
	Lb.	Lb.	Lb.	Lb.	Lb.
Nitrogen .	50	49	98	149	49
Soda . .	2·6	5·0	32·0	118·7	9·2
Potash . .	28·8	35·7	79·7	300·7	50·9
Magnesia .	7·1	6·9	9·2	42·5	14·4
Lime . .	9·2	9·2	42·4	42·9	32·1
Phosphoric Acid	21·1	20·7	21·7	52·9	12·3
Sulphur . .	7·8	6·1	17·8	14·0	5·7
Chlorine .	2·5	4·1	15·1	83·1	14·6
Silica . .	96·9	68·6	6·7	17·9	56·9

plant food the production may shrink in a very few years to one-third or one-fourth of the average full crop. Once, however, the yield has reached this lower level, it will remain for an indefinite period comparatively stationary, affected only by the fluctuations due to season. At Rothamsted, for example, wheat has now been grown year after year on the same land for a century, and one plot has received no manure throughout the whole period. In the first few years the crop declined steadily, but since then little or no further drop has been seen. The yield remains at about 12 bushels per acre for each successive ten years' average, and has considerably exceeded that amount during favourable seasons. This yield, however, of 12 bushels of corn per acre, is only about a third of that obtained on the adjacent plots receiving manure every year during the same period.

These facts lead to the conclusion that it is not merely the total amount of this or that plant food in the soil which is to be considered but rather the form in which it is present. The material may be soluble in the strong acid used in the analysis, but yet quite beyond the reach of the plant. The plant can of course make use of any substances which have been brought into solution in the soil and hence it becomes necessary to discriminate between the available and non-available plant food.

Accepting, then, the fact that the soil contains a vast store of all the elements necessary to its nutrition, it remains to ascertain which of the substances are normally likely to fall below the current requirements of the crop. This is a question that can be solved only by field experiments, and though the answer will vary with each crop and each soil, yet certain general principles at once become evident and upon them the whole idea of a fertilizer is based. For example, field experiments clearly show that, as a rule, certain elements indispensable to the plant need not be supplied to the crop in the field, since the normal soil is able to provide a sufficiency. Calcium, magnesium, iron and sulphur fall into this class. But there are three elements required for the nutrition of the crop—

nitrogen, phosphorus, and potassium—that soils cannot always supply in sufficient quantities or rapidly enough. Fertilizers, then, are designed to supply deficiencies in the soil, and for all practical purposes are to be regarded as consisting of compounds of nitrogen, phosphorus, and potassium, either singly or together. They may also contain magnesium, calcium, or sulphur, but these, though equally necessary to the plant, are not taken into consideration in the sale of fertilizers.

To sum up, then, we find that the plant must obtain the elements carbon, hydrogen, oxygen, nitrogen, sulphur, phosphorus, calcium, magnesium and potassium in substantial amounts and several others such as iron, manganese, copper, zinc and boron in smaller amounts. The first three are provided by the atmosphere and water, while the remainder must come from the soil. Normal soils generally have abundant reserves of these elements but they are not always in a form which can be absorbed by the plant; sometimes these reserves have been exhausted by continued cropping or by leaching. The three elements which are most commonly deficient in available form in soils are nitrogen, phosphorus and potassium, and these deficiencies are made good by adding fertilizers in amounts related to the quantities required for full crops. Calcium deficiency is also a very common cause of crop failure but only where the sound practice of regular liming has been neglected.

A distinction is often drawn between "natural" and "artificial" manures. Properly speaking the latter term should include all such materials as result from manufacturing processes, the former term being reserved for waste vegetable and animal products. Ammonium sulphate and superphosphate undoubtedly belong to the category "artificial," whilst guano and decomposing straw belong to the "natural" class. But it is very difficult to draw a precise line separating the two types. Inorganic material like Chilean nitrate and potassium salts occur naturally, but are put through processes of purification before they are marketed, whilst many organic manures such as bone meal and spent hops are by-products of industrial processes; even farmyard manure is not, strictly speaking, a natural product. Perhaps the best distinction is that between bulky organic substances on the one hand and the more concentrated and mainly inorganic compounds on the other, and there is a growing tendency at the present time to refer to "organic manures" and "inorganic" or "mineral fertilizers." But, as indicated at the beginning of this chapter, the terms manure and fertilizer will probably long continue to be regarded as interchangeable.

CHAPTER II

THE SOURCES OF NITROGEN IN VEGETATION AND SOIL

Need for combined nitrogen—Nitrogen-fixing organisms—Fertility level.

Need for Combined Nitrogen

AMONGST the elements of nutrition of the plant the first place must be given to nitrogen. Not only does it cost more than the other necessary elements, but as a fertilizer applied to ordinary soils it usually has a more direct and immediate effect upon the plant; furthermore, it differs from the others in that plants live habitually in contact with a vast unusable store of it. Since plants live in an atmosphere four-fifths of which consists of elementary nitrogen, it is perhaps necessary to justify a little the statement made in the previous chapter, that they obtain the nitrogen they require only in a combined form and by means of their roots. The form that the demonstration has taken may be seen in the water culture experiment which has already been illustrated; in the absence of combined nitrogen, the development of the plant is very small. The same is true for cultures in sand, which reproduce more closely the natural conditions, and many careful experiments have been performed with plants growing in such artificial soils supplied with a known amount of nitrogen. When the plants have come to the full term of their growth, the nitrogen they contain is found to be exactly balanced by the amount of the same element which has been removed from the soil. Among these experiments, a most elaborate series were carried out at Rothamsted in 1857-58, and were generally regarded as definitely settling the question against the fixation of nitrogen by the plant itself.

The experiments were made with wheat, barley, oats, clover, beans, peas, and buckwheat, and the trials were repeated, in one case with no manure in the pots, and in another with the supply of a small quantity of ammonium sulphate. The soils employed were made up from either ignited pumice or ignited soil, and the glass shades under which the plants were grown rested in the groove of a stoneware vessel, mercury being used as a seal. The air, previously passed through sulphuric acid and sodium carbonate solution and washed, was forced into the apparatus, so as always to maintain a greater pressure inside than out, thus minimizing all danger of unwashed air leaking in; carbon dioxide was also introduced as required. Under these rigorous conditions the results in Table 3 were obtained.

And if objection be made that such plants were enfeebled by the unnatural conditions, so that they had lost their power to bring nitrogen into combination—to "fix" it, in current language—there are many other types of experiment which render such criticism invalid. For example, Hellriegel performed a long series of experiments with different plants, which showed, up to a point, that the amount of growth was very closely proportional to the amount of nitrogen supplied in a combined form, when there was a sufficiency of the other elements of plant food present. This would not be the case were the plant able to get any nitrogen for itself from the atmosphere.

Again, to meet an early objection of Liebig and his followers that the Rothamsted crops failed to draw upon the nitrogen of the air because they had not the necessary initial development

TABLE 3.—SUMMARY OF THE RESULTS OF EXPERIMENTS MADE AT ROTHAMSTED TO DETERMINE WHETHER PLANTS ASSIMILATE FREE NITROGEN.

			Nitrogen.—Gram.			Ratio of Nitrogen recovered to Nitrogen supplied.
			In Seed and Manure if any.	In Plants, Pot, and Soil.	Gain or Loss.	
WITH NO COMBINED NITROGEN SUPPLIED BEYOND THAT IN THE SEED SHOWN.						
Gramineæ .	1857	Wheat .	0·0080	0·0072	—0·0008	0·90
		Barley .	0·0056	0·0072	+0·0016	1·29
		Barley .	0·0056	0·0082	+0·0026	1·46
	1858	Wheat .	0·0078	0·0081	+0·0003	1·04
		Barley .	0·0057	0·0058	+0·0001	1·02
		Oats .	0·0063	0·0056	—0·0007	0·89
	1858	Wheat .	0·0078	0·0078	...	1·00
		Oats .	0·0064	0·0063	—0·0001	0·98
Leguminosæ	1857	Beans .	0·0796	0·0791	—0·0005	0·99
	1858	Beans .	0·0750	0·0757	+0·0007	1·01
		Peas .	0·0188	0·0167	—0·0021	0·89
Other Plants .	1858	Buckwheat	0·0200	0·0182	—0·0018	0·91
WITH COMBINED NITROGEN SUPPLIED.						
Gramineæ .	1857	Wheat .	0·0329	0·0383	+0·0054	1·16
		Wheat .	0·0329	0·0331	+0·0002	1·01
		Barley .	0·0326	0·0328	+0·0002	1·01
		Barley .	0·0268	0·0337	+0·0069	1·25
	1858	Wheat .	0·0548	0·0536	—0·0012	0·98
		Barley .	0·0496	0·0464	—0·0032	0·94
		Oats .	0·0312	0·0216	—0·0096	0·69
	1858	Wheat .	0·0268	0·0274	+0·0006	1·02
		Barley .	0·0257	0·0242	—0·0015	0·94
		Oats .	0·0260	0·0198	—0·0062	0·76
Leguminosæ	1858	Peas .	0·0227	0·0211	—0·0016	0·93
		Clover .	0·0712	0·0665	—0·0047	0·93
	1858	Beans .	0·0711	0·0655	—0·0056	0·92
Other Plants	1858	Buckwheat	0·0308	0·0292	—0·0016	0·95

of leaf, one experimental plot there was supplied with a very small amount of active nitrogenous manure, just to give the young plant a good start, so that it might be able to continue to feed upon the atmospheric nitrogen. But, as

TABLE 4.—ROTHAMSTED MANGOLDS (1876–1902).

	Roots per Acre.	Increase per lb. of N.
	Tons.	Tons.
Superphosphate and Sulphate of Potash . .	4·55	...
Do. do. + 7·8 lb. N. .	5·93	0·17
Do. do. + 86 " .	14·03	0·11
Do. do. + 93·8 " .	14·60	0·107

Table 4 shows, the small addition of nitrogen produced only a small increase of crop compared with the much larger increase produced by a normal application of the same fertilizer. If, then, the yield of most of our field crops is, until some other limiting factor comes into play, almost proportional to the amount of combined nitrogen they receive, it is reasonable to conclude that they have drawn none from the atmosphere. It is indeed true that the atmosphere does contribute a small amount of nitrogen for the use of the plant under ordinary conditions, because traces of both ammonia and nitric acid are found in the air and are washed out by the falling rain. Table 5 shows the average amount of nitrogen as nitric acid and ammonia brought down in the rain falling at Rothamsted for the thirteen years between 1st September, 1888, and 30th August, 1901, together with the corresponding results obtained at a few other places where observations have been made for any long period. It will be seen that the Rothamsted results are considerably lower than those obtained at the Paris, Copenhagen, or Florence stations, though they do not differ notably, as regards the total amount of nitrogen falling per acre, from those obtained at the two tropical stations in the West Indies. The

TABLE 5.—NITROGEN AS AMMONIA AND NITRIC ACID IN RAIN.

Locality.	Date.	Rain-fall.	Nitrogen. Per Million. As Ammonia.	Nitrogen. Per Million. As Nitric Acid.	Nitrogen. Lb. per Acre per annum. As Ammonia.	Nitrogen. Lb. per Acre per annum. As Nitric Acid.	Nitrogen. Lb. per Acre per annum. Total.
Rothamsted	1888–1901	27·25	0·440	0·183	2·71	1·13	3·84
Copenhagen	1880–1885	21·95	1·97	0·473	9·27	2·21	11·48
Montsouris	1876–1900	21·52	2·13	0·66	10·37	3·22	13·59
Florence	1869–1875	38·31	1·004	0·57	8·70	3·09	11·79
Barbados	1885–1897	63·95	0·084	0·268	1·22	3·88	5·10
British Guiana	1890–1900	102·41	0·055	0·078	1·17	1·82	2·99

high results are probably due to the proximity of towns, because the majority of other determinations, not quoted here, agree more nearly with the Rothamsted figures. It may thus be assumed that ordinary land receives about 4 to 5 lb. per acre per annum of combined nitrogen from the atmosphere, an amount which forms only a small fraction of the requirements of the crop.

The tenacity with which in the face of such evidence the opinion was held that the leaf of the plant could obtain nitrogen as well as carbon from the atmosphere, was due to the difficulty of explaining how the world's original stock of combined nitrogen could have arisen. Assuming the world to have cooled down from the state of incandescent gas, it must have started with all its nitrogen in the free gaseous state; yet as we see it to-day, the whole stock of combined nitrogen is of organic origin.

Nitrogen-fixing Organisms

The circulatory process through which combined nitrogen passes is quite plain. Animals consume in their food the very complex nitrogenous substances called proteins; these are broken down in the alimentary canal to simpler substances by the digestive processes; the by-products in animal nutrition —urea and comparatively simple derivatives of ammonia—are excreted along with undigested materials. In due course, the excreta undergo decomposition in the soil to ammonium salts and nitrates which are absorbed by the plant and once more built up into proteins. The stocks of combined nitrogen that have been handed down from past ages all speak of the same organic cycle, never of fixation. Coal is but the debris of an extinct vegetation; nitrate beds are the result of some process of decay. Virgin soils may acquire vast stores of nitrogenous humus through long periods of vegetable growth; but if plants cannot fix nitrogen, there can have been no gain, however long the growth, but only a circulation of the pre-existing combined stock. Inorganic agencies are not important ones because nitrogen gas is a very inert substance and difficult to bring into combination with other elements. It is true that it will combine with oxygen under the influence of an electric spark so that lightning flashes may account for nitric acid in the atmosphere, but Table 5 shows that the amount in rain water is extremely small.

However, one vital process, which has probably been operative since the beginning of life on the earth, lies in the power of certain bacteria to assimilate elementary nitrogen. Though the demonstration in field and laboratory experiments, that plants can make use of nitrogen only when it is in some form of combination, seemed beyond criticism, there was one group of plants—peas, beans and clover—which derived little or no benefit from nitrogenous fertilizers. These crops removed an exceptionally large amount of nitrogen from the soil and yet left it richer in nitrogen than it had been before their growth.

That beans or vetches or lupins were the best preparation for a wheat crop was a commonplace of Roman agriculture, and the same observation afterwards became enshrined in that most fundamental of rotations, the Norfolk four-course system, in which wheat follows clover or beans. Hellriegel and Wilfarth found that leguminous plants did gather nitrogen from the atmosphere, and could, therefore, become wholly independent of nitrogenous manures; but this only took place when, by infection from the soil, certain characteristic nodules were formed upon the roots. These nodules were found to be colonies of a particular bacterium (*Rhyzobium*) which lives symbiotically on the host plant, furnishing it with nitrogenous matter and deriving from it the carbohydrate required for the fixation of nitrogen. The fixation of nitrogen is a chemical process which requires a supply of energy from outside, and this the bacteria obtain by the oxidation of carbohydrate.

The leguminous plants therefore form, by their association with nitrogen-fixing bacteria, one considerable natural source of combined nitrogen, and how effective they can be in accumulating fertilizing matter in the soil may be judged from the accompanying Table 6, showing the results of one of the Rothamsted experiments upon crops grown in rotation.

On this field (Agdell) the rotation begins with a crop of

TABLE 6.—EFFECT OF CLOVER ON SUCCEEDING CROPS.

Manuring for Swede Crop only.	Clover, 1894	Wheat, 1895.			Roots, 1896.			Barley, 1897.		
		After Fallow.	After Clover.	Increase due to Clover.	After Fallow.	After Clover.	Increase due to Clover.	After Fallow.	After Clover.	Increase due to Clover.
	Cwt.	Lb.	Lb.	Per cent.	Cwt.	Cwt.	Per cent.	Lb.	Lb.	Per cent.
Mineral manure	59·7	4,220	5,180	+22·7	179·1	244·5	+36·5	2,103	3,991	+89·8
Complete manure	76·7	4,547	5,209	+14·6	379·8	388·8	+2·4	3,595	4,913	+36·7

Swede turnips, which is manured, in one case with mineral manures, in the other with a complete fertilizer. Following the swedes comes barley without manure, then the field is divided, and on one portion clover is grown, while the other is bare fallowed and carries no crop throughout the year. A crop of wheat, still unmanured, completes the rotation. In the table the yield on the portions which have grown clover is compared with that on the portions without crop; it will be seen that although a crop of nearly three tons of clover hay has been taken away from the one portion, the wheat which follows it is 23 per cent. better than on the portion where no clover had been grown in the previous year. Nor is the benefit due to the clover, exhausted by the wheat crop, for it is seen to persist in the root crop following the wheat and in the barley which comes a year later still.

The nitrogen-fixing bacteria living in symbiosis with leguminous plants are highly specialized, and their efficiency is very variable. The clover bacteria in some poor hill pastures, for example, are ineffective. But effective strains have been isolated, which can compete successfully with the ineffective indigenous strains, and so a simple method of improving the growth of legumes is provided by inoculation. This has met with conspicuous success in the introduction of lucerne into new districts in which it could not be established previously. The seeds are treated with a pure culture of an appropriate strain of the nodule-forming organism and, provided the soil conditions with respect to lime and phosphate are suitable for the growth of these crops, the seedling roots become inoculated and the plants develop vigorously. And there is no reason to suppose that this work with useful bacteria will not expand and enhance this very valuable process by which nitrogen is brought from the atmosphere into an organic form in the soil.

The only practical limitation to the gathering of nitrogen by this method lies in the difficulty that is found in growing some leguminous crops frequently on the same land. Although the Rothamsted experiments have demonstrated that it is possible to grow wheat year after year for a century and maintain the yield if the appropriate manures are employed, on many soils red clover cannot be grown with success more frequently than once in 5 or 6 years. As the farmer says, the land becomes "clover sick," and though the clover seed germinates and grows for a time, the constitution of the plant is so weak that it almost inevitably succumbs during the winter. The cause of clover sickness is attack either by a parasitic stem eelworm, or by a fungus *Sclerotinia trifoliorum*. It is controlled by substituting crops less susceptible than clover or by lengthening the rotation.

Other bacteria have been discovered in the soil which are capable of fixing free atmospheric nitrogen without association with any host plant, provided they are supplied with some carbohydrate, by the oxidation of which they derive the energy necessary to bring the nitrogen into combination. Of these bacteria the best known and probably the most effective is a large organism, discovered by Beijerinck in Holland, and called by him *Azotobacter chroococcum*.

Similar bacteria have been found in cultivated soils all over the world. The organism is extremely sensitive to acidity and does not occur in soils whose pH value is under 6. Roughly speaking, its action is to oxidize carbohydrates to carbon dioxide and water, forming as by-products certain organic acids and some dark brown humus (whence the name "chroococcum") and at the same time bringing atmospheric nitrogen into combination. Under the most favourable conditions, however, nitrogen is fixed to the extent of only 1 or 2 per cent. of the carbohydrate consumed and it is doubtful whether this could account for the accumulation of nitrogen in the rich black soils of the Russian Steppes, or of the American Middle West. Certainly, as long as these lands were uncultivated, the annual return of vegetable matter furnished conditions suitable for Azotobacter, the carbohydrates providing the organism with its necessary food supply; but less importance is attached to the activities of Azotobacter now than formerly. When the crop is removed from cultivated soils there is a fairly rapid drop in the nitrogen of the soil. This may be due in part to the fact that there is no longer much chance of accumulation, but it is mainly due to loss of gaseous nitrogen by certain decomposition processes.

When land is laid down to grass there is a recovery in amount of combined nitrogen in the soil. A good example of this is found in two pieces of land at Rothamsted, which for the last fifty years have been allowed to run wild and assume a natural prairie condition of self-sown weeds and grasses, that are never taken away but left to rot where they die down. Samples of the soil had been taken at the beginning of the period, and by comparing them with more recently taken samples it has been possible to detect a very considerable fixation of nitrogen, amounting in the most favourable case to nearly 50 lb. nitrogen per acre per annum. The second similar piece of land shows a much lower result, and this is correlated with a greater degree of acidity in the soil of this plot.

It is too early yet to speculate freely on the work of the various nitrogen-fixing bacteria; we may, however, confidently attribute to their action both the current stock of combined nitrogen in the world and the continued source of its renewal in the future.

Fertility Level

These natural processes for bringing nitrogen into combination and so regenerating the stock of combined nitrogen which is always undergoing a certain amount of waste, cannot act rapidly enough to maintain the soil at a high level of fertility. When they have been in operation for a long period of time there may be accumulated such a stock of combined nitrogen that the soil for many years will produce a maximum output without any further addition of nitrogen. Such was the case with the black soils of the Middle West both in the United States and Canada, and also with the black soils of the English fens; but crop production under these conditions resembles mining, in that a stock that is finite, however great at the outset, is being exploited and no regeneration is taking place, so that in the end the high production begins to decline. Eventually, indeed from the beginning on more normal soils, the level of fertility that can be maintained by bacterial fixation of nitrogen is not high.

For example, the British system of farming prior to 1840 was based upon the return to the land of all the fertilizing constituents of the crops grown, except such as were contained in the wheat and barley, meat and milk produced, and on the fixation of a certain amount of nitrogen by the growth of a clover or bean crop once in every four years. The natural regeneration of the nitrogen supply thus obtained was sufficient to maintain indefinitely the crop production of wheat, barley, turnips, and hay at a level measured by a wheat production of about 20 bushels per acre. To go beyond that some extraneous source of nitrogen in fertilizers was necessary. Crops at a much higher level are produced by growing long leys rich in wild white clover of three, four, or five years in duration, then ploughing up and growing three or four crops on the fertility accumulated. This practice of "ley farming" has been considerably extended with promising results but, of course, the higher production has to be spaced over a longer period. An increased annual production can only be assured by bringing in some nitrogen from an outside source, i.e. a fertilizer that represents some natural but exhaustible stock or one that has been produced by a manufacturing process which is capable of bringing nitrogen into combination.

CHAPTER III

FERTILIZERS CONTAINING NITROGEN

Sodium nitrate—Potassium nitrate—By-product ammonium sulphate—World production and consumption of nitrogen—Fixation of atmospheric nitrogen—Calcium cyanamide—Calcium nitrate—Synthetic ammonium sulphate—Production costs—Urea—Waste products as nitrogenous manures, soot, shoddies, dried blood.

THE most important compounds discussed in this chapter and the percentage of nitrogen usually guaranteed in the sale of each, are as follows:

Sodium Nitrate (Chilean Nitrate of Soda)	16·0
Ammonium Sulphate (Sulphate of Ammonia) . .	20·6
Calcium Cyanamide, commercial	20·6
Calcium Nitrate (Nitrate of Lime)	13–16
Ammonium Nitrate	35·0
Nitrochalk (Ammonium Nitrate and Calcium carbonate) .	15·5
Urea	46·0

Sodium Nitrate

Prior to 1910 the two main sources of nitrogenous fertilizers were sulphate of ammonia (ammonium sulphate), obtained as a by-product in the combustion of coal, and nitrate of soda (sodium nitrate), imported from Chile. Nitre or saltpetre (potassium nitrate), which is produced in India by boiling wood ashes with nitrate-rich soil extracts, had long been known to possess great manurial value. Consequently, when the existence of great deposits of nitrate was discovered in Chile at the beginning of the nineteenth century, the question of exporting the material to Europe was soon taken up. By 1840 it had been tried experimentally by many farmers, and the production and consumption grew rapidly to reach the figure of 2·87 million tons per annum just before the war of 1914–18. The output actually rose to 3 million tons in 1917 to meet the heavy requirements by the Allies for explosives and, after post-war fluctuations, due to various causes, had become stabilized at about 1·5 million tons previous to the outbreak of the 1939–45 world war. Incidentally, iodine is a valuable by-product of Chilean nitrate which accounts for about three-quarters of the world's output.

The Chilean deposits occur in irregular formations extending over some 75,000 square miles of desert lying between the coastal hills on the Pacific Coast and the Andes. Only a very small proportion of this area has been fully explored, so that estimates of the possible period of exhaustion are largely speculative. Furthermore, improved methods of extraction can deal with raw material which previously could not be worked economically. The deposits lie about 3,000 feet above sea-level on a plateau of an average width of 35 miles. The climate is intensely dry, rain falling only every two or three years and then in such small quantities as evaporate rapidly.

The nitrate-bearing deposit, or *caliche*, occurs a few feet under beds of sand and gravel and clay, and contains many salts in addition to sodium nitrate, of which there may be from 75 down to 7 or 8 per cent. One theory regarding the origin of the deposits is that the plain is an ancient sea-bed elevated by volcanic movements and then desiccated; the oxidation of immense masses of seaweed originally present in the sea is regarded as the source of the sodium nitrate. The presence of iodine in the *caliche* is one argument in support of this theory because seaweed contains iodine, but there are many arguments against it. For example, seaweed also contains bromine but there is no bromine in the *caliche*. More conclusive is the fact that, from our knowledge of the nitrification process, it is not possible to conceive of its taking place in a rapidly-concentrating solution of common salt. A much more reasonable theory is that the nitrates and other salts were derived from rich soils or rocks on the high ground of the Cordilleras, and transferred to the lower slopes through the agency of atmospheric moisture. According to this hypothesis, the salts were subjected to a continual process of solution in rain or fog and redeposition as the moisture evaporated, coming nearer and nearer at each stage to the point where they are found to-day.

The only other deposits of sodium nitrate which are known to have had some economic importance are those occurring in Upper Egypt. Certain shale beds of Eocene age, outcropping on both sides of the Nile, contain about 6 or 7 per cent. sodium nitrate, associated with sodium chloride and sodium sulphate, which at one time made the clay or "tafla" worth carriage in the vicinity, as a manure. Analyses indicate that the nitrate was formed *in situ* from nitrification of a mass of organic matter originally contained in the deposit, but further data are required to provide a clear picture of the occurrence. This much can be said, however; the Egyptian deposits suggest that the possible formation of the Chilean beds is due to the leaching of salts from similar strata, the

Cordilleras consisting of rocks of recent age, into a rainless area and their accumulation by evaporation. One feature of both deposits is the relatively large proportion of sulphates which seems to dispose of any question of marine origin, but they present the common difficulty that the nitrate concerned is sodium nitrate, whereas the usual product of nitrification in the soil is calcium nitrate. However, under arid conditions sodium salts are characteristically abundant, and sodium carbonate, which is always present, would form a base for nitrification. At the same time, the sulphur of the organic matter would be oxidized to sulphate. But, to settle the problem of the origin of the Chilean deposits and to account for the presence of iodine, a full examination is required of the rocks of the Cordilleras, the drainage from which would find its way into the nitrate plateau.

In the *caliche*, the sodium nitrate is associated with sodium chloride and sulphates of calcium, magnesium, sodium and potassium, and a varying amount of earthy material. The deposit is broken up by blasting, removed to the works and lixiviated in large vats with water at 60°C. The nitrate is much less soluble in cold than in hot water, so the saturated solution is cooled by aid of refrigeration to 10°C and the nitrate crystallizes out and is separated from the mother liquor. It is ready for use after draining and drying. This is the older form of Chilean nitrate and is a coarse crystalline salt, greyish-white in colour but sometimes with a pink or brown tint. The newer form is granulated and consists of small round white pellets which are obtained by fusing the crystalline material and spraying the molten material into a spacious cooling chamber; it keeps condition better than the crystals and is very easy and pleasant to handle. Chilean nitrate is now shipped in bulk as well as in bags. It is about 96 per cent. pure sodium nitrate and contains 16 per cent. of nitrogen; a typical specimen might show the following composition:

Sodium Nitrate	96·0
Potassium Nitrate	1·4
Moisture	1·6

the remaining 1 per cent. consisting principally of chlorides and sulphates of sodium, magnesium and calcium, sodium iodate and sodium borate. In addition there are usually traces of a great number of other elements. Sodium nitrate is produced synthetically in some countries and differs from the salt which occurs naturally by the absence of these many other substances. It is possible that these impurities are of value in helping to correct deficiencies of trace elements in certain soils.

The salt is very soluble in water and deliquescent so that, when exposed to damp conditions, it may absorb enough water to become very sticky and even go into solution. Hence it should be stored in a dry place. Nitrates in quantity are poisonous so precautions should be taken against stock obtaining access to the fertilizer or to unwashed bags. The dry salt, on account of its high percentage of oxygen, accelerates the combustion of such things as wood and clothing so it is advisable to avoid bringing materials which have become saturated with the salt into proximity of a flame. Like all soluble salts, it can scorch by plasmolysis any green tissue with which it is left in contact, and consequently it must be sown with some care when used as a top dressing on a growing crop like cabbage. It is particularly valuable as a top dressing when plants are backward for some reason, because its effect is so rapid. It is very soluble and hence the plant soon has access to nitrate in solution and generally responds in two or three days. To avoid unnecessary loss of nitrate to the drainage, it is usually applied to a growing crop, but in some cases, e.g. sugar beet, part of the dressing may be applied at seeding.

In the season 1938–39, the world consumption of Chilean nitrate was 1·55 million tons, of which Europe took about half and U.S.A. about one-third. Approximately 45,000 tons were consumed in the British Isles. Outside the northern temperate climate Egypt was the largest consumer with 170,000 tons.

Potassium Nitrate

In the majority of cases, the *caliche* contains 2 or 3 per cent. of potassium nitrate, but in some ores the figure may reach 5 to 7 per cent. A small quantity of potassium nitrate is invariably present in commercial Chilean nitrate; but it is obviously an advantage to modify the usual processing so as to obtain a mixture rich in potassium, and a product called Chilean Potash Nitrate has been marketed. This now contains roughly 20 per cent. potassium nitrate and, on the assumption that the bulk of the remaining 80 per cent. is sodium nitrate, will show about 10 per cent. K_2O and 15 per cent. N on analysis. Three methods have been employed to obtain salts rich in potassium nitrate; they are based upon the different solubilities of the sodium and potassium salts at different temperatures. Fractional crystallization of the hot liquor from the tanks is the simplest and cheapest means, but is practicable only when the ore contains about 5 per cent. potassium nitrate. Evaporation of the mother liquor, after separation of the ordinary nitrate, concentrates the various salts present and the concentrated solution is returned to the boiling tanks to reduce the quantity of impurities before subsequent fractionation to produce a potassium-rich mixture.

This evaporation procedure has been largely displaced, however, by one involving refrigeration of the mother liquor. This method is based upon the fact that, at low temperatures, the solubility of potassium nitrate decreases more rapidly than that of sodium nitrate so that the crystalline deposit becomes greatly enriched in the former; further, the solubility of boric acid, magnesium sulphate and sodium chloride are unaffected so that these substances are not deposited. Hence a much purer product is obtained than by straightforward evaporation, containing as much as 98 per cent. of nitrates of which, as stated above, 40 per cent. may consist of potassium nitrate. About 60,000 tons of this nitrate mixture were produced in 1938 and it obviously possesses distinctive merits as a fairly concentrated and readily soluble fertilizer containing potassium to balance the nitrogen.

Ammonium Sulphate (by-product)

Coal contains between 1 and 2 per cent. of nitrogen, derived from the original vegetable matter from which it has been formed, and when coal is burned some of this nitrogen is given off as ammonia. Ammonia is therefore a valuable by-product of gas-works, coke ovens, blast furnaces and also of shale oil distillation plants. In the modern gas works, about 35 to 40 per cent. of the nitrogen is found in the tar and gases, the remainder being left in the coke. The ammonia is recovered from the gases by scrubbing and may be distilled off from the resulting ammoniacal liquor into sulphuric acid. The ammonium sulphate formed is then separated by concentration of the solution and crystallization. One ton of coal yields about 22 or 23 lb. of ammonium sulphate, which is equivalent to about a fifth of the nitrogen in the coal.

The pure compound, ammonium sulphate, is a white crystalline salt easily soluble in water; the commercial by-product varies somewhat in colour due to traces of tarry matter, being generally grey or yellow, but sometimes it is bluish owing to the presence of a little ferrocyanide derived from the cyanides present in the coal gas. Occasionally a sample may contain some ammonium thiocyanate which is toxic to vegetation, but it is most unusual to find this impurity in harmful amounts. Like all ammonium salts, the sulphate reacts with basic substances like lime, calcium carbonate and basic slag, giving off ammonia as a gas. To prevent loss of nitrogen, therefore, it should not be mixed with these substances. The pure salt contains 21·2 per cent. nitrogen; the commercial product is usually guaranteed to contain 20·6 per cent.

By-product sulphate of ammonia was a profitable undertaking and contributed 200,000 tons of nitrogen or about a third of the total world supply of combined nitrogen in 1910. In 1938 the production had reached 400,000 tons per annum, but accounted for less than one-sixth of the total world production of nitrogen. The reason was that there had been a great increase in the demand for nitrogenous compounds since the beginning of the century, and a spectacular development in the synthetic processes of manufacturing them.

The following figures illustrate the remarkable change in the total and relative amounts of inorganic nitrogen products during the years 1908 to 1953, all the figures being in terms of nitrogen.

TABLE 7.—WORLD PRODUCTION AND AGRICULTURAL CONSUMPTION OF NITROGEN.
(in thousands of metric tons of 2,205 lb.)

Year.	1908.	1937–38.	1952–53.
Chilean Nitrate	307	224	270
Ammonium Sulphate—by-product	161	411	480
—synthetic	...	765	1,523
Calcium Cyanamide	1·0	305	388
Calcium Nitrate	1·0	195	290
Other Forms of Nitrogen*	2·0	980	2,314
Agricultural Consumption about	472	2,492	5,265
Total Production	490	2,880	6,187

The most striking feature in the above table is the very great increase in the consumption of nitrogenous fertilizers, corresponding in the main to the remarkable development of the synthetic nitrogen industry. Of the total production in 1952–53, amounting to 6·19 million metric tons of nitrogen, 2·80 million tons were consumed in Europe, including U.S.S.R., 0·17 in Africa, 0·93 in Asia, 0·02 in Oceania, and 2·11 in the Americas. Of the world total agricultural con-

* This includes various nitrogenous compounds, synthetic and by-products of industry, used as nitrogenous fertilizers like ammonia, ammonium phosphate, ammonium chloride, ammonium nitrate, urea and "concentrated complete fertilizers." Nitrogen compounds in solution now account for a large proportion of the total fertilizer nitrogen in the U.S.A.

sumption in 1952–53 of 5·3 million tons, Chilean nitrate accounted for about 5 per cent., ammonium sulphate for nearly 33 per cent., ammonium nitrate for 24 per cent., and the other chief synthetic fertilizers for about 30 per cent.

In the United Kingdom the total production of nitrogen compounds expressed as ammonium sulphate reached 2 million tons in 1953; the agricultural consumption of nitrogen was about 136,000 tons in ammonium sulphate, and 95,000 tons in other forms; the total agricultural consumption in 1952–53 was 231,000 metric tons of nitrogen. This consumption amounted to about 17 lb. nitrogen per acre of agricultural land (exclusive of rough grazing and forest); in contrast, W. Germany was using approximately 23 lb., Belgium 42 lb., and Netherlands 71 lb. per acre.

The potential nitrogen fixation capacity of industrial plants has increased in many countries in recent years and there would seem to be little danger of failure to meet an increasing agricultural consumption in the near future.

Fixation of Atmospheric Nitrogen

Of the many laboratory methods by which nitrogen may be brought into combination, only three have proved suitable for commercial development on a large scale. At extremely high temperatures, such as are obtained in an electric arc, nitrogen will combine with oxygen to form various oxides from which nitric acid will eventually result by interaction with air and water. Again, if a mixture of hydrogen and nitrogen is heated (in the stock class-room experiment by passing sparks through it) a certain small amount of ammonia is formed and the extent to which combination takes place is conditioned by pressure, temperature, and other factors. Lastly, nitrogen will combine with a few metals and compounds, again at high temperatures, to yield substances which under the action of water give rise to ammonia. It was this last method which was first developed on a commercial scale by Frank and Caro in Berlin.

Calcium Cyanamide

They started with calcium carbide, the substance so well known as the source of acetylene for heat and light. Frank and Caro found that this compound would combine readily with nitrogen gas at quite moderate temperatures, and that the resulting substance, calcium cyanamide, would decompose under the action of water, yielding its nitrogen as ammonia and the calcium and carbon as calcium carbonate. An Italian company, which was the first to take up the patents for the manufacture of calcium cyanamide, established its factory alongside one of the great producers of calcium carbide at Piano d'Orte in the hills above Rome, where water-power can be obtained for the cheap generation of electricity. Then other works were erected in Norway, in Savoy, and in America, where suitable water-power was available. Carbide was not manufactured in this country until a large-scale modern plant was established near Swansea during the war. Production is said to be about 100,000 tons per annum, but mainly for acetylene burners. In the manufacturing process, the calcium carbide is first roughly ground and then heated in steel drums through which a current of nitrogen gas is passed. The calcium carbide, which itself results from the reaction of a mixture of lime and coke in the electric furnace, must either be purchased or manufactured by a preliminary process. The two reactions of forming the carbide and uniting it with nitrogen can indeed be carried out simultaneously, but this method has been abandoned in practice. The nitrogen gas was originally obtained by passing a current of air over red-hot copper, the copper oxide formed being afterwards reduced to the metallic state again by sending over it, while still hot, a current of coal-gas. More recently a process of obtaining nitrogen by fractional distillation from liquefied air has been employed, but in this case there must also exist a market for the compressed oxygen.

The carbide is heated electrically to about 800°C by a carbon rod inside the drum, and a mixture of calcium cyanamide and carbon is produced, the reaction giving off heat which raises the temperature to 1100°C.

$$\underset{\text{calcium carbide}}{CaC_2} + N_2 = \underset{\text{calcium cyanamide}}{CaCN_2} + \underset{\text{carbon}}{C}$$

The process is allowed to proceed for 36 hours, cooling takes a further 24 hours, then the material is broken up in a ball mill to give a dark grey powder which is moistened to decompose any free carbide remaining.

Pure calcium cyanamide contains 35 per cent. nitrogen, but the fertilizer, (often unfortunately called "cyanamide"), contains only about 60 per cent. calcium cyanamide together with 10–12 per cent. carbon, about 20 per cent. free lime and small quantities of many other substances. Commercial calcium cyanamide usually carries a guarantee of 20·6 per cent. nitrogen. At one time it was very dusty and unpleasant to handle, but special processes were devised to coat the particles with a small amount of vegetable or mineral oil. A further improvement has been the production of a granulated form; one method of achieving this is to add sufficient water to hydrate the free lime and then to mix with calcium nitrate and set aside to harden; the product is finally crushed and screened. This granular calcium cyanamide thus contains from 1·5 to 3 per cent. nitrogen as nitrate which, unlike the bulk of the nitrogen, is immediately available to the plant.

Calcium cyanamide behaves quite differently from the more common nitrogenous fertilizers which consist of ammonium salts or nitrates, for it is actually toxic to young plants, being commonly employed as a weed killer, and is harmful to the nitrifying organisms in the soil. The reason is that it undergoes a somewhat complicated series of chemical changes which may be conveniently summarized in the following manner. It is slightly soluble in water and decomposes to give acid calcium cyanamide and calcium hydroxide.

$$2CaCN_2 + 2H_2O = Ca(HCN_2)_2 + Ca(OH)_2$$

In presence of an acid, cyanamide (H_2CN_2) is formed.

For example, carbon dioxide and water (carbonic acid) give rise to cyanamide and calcium carbonate, a change which readily takes place in a damp atmosphere or in the soil.

$$CaCN_2 + H_2O + CO_2 = H_2CN_2 + CaCO_3$$

The cyanamide is then further broken down to give urea.

$$H_2CN_2 + H_2O = CO(NH_2)_2$$

This last process proceeds very slowly in solution but can take place in 2 or 3 days in the soil for it seems to be catalyzed by certain colloidal particles. Urea is quite rapidly converted by the action of soil bacteria into ammonium carbonate which is in turn oxidized, but relatively slowly, to nitrate.

However, cyanamide is liable to undergo another change and form a polymer called dicyanodiamide, $H_4C_2N_4$, especially under the influence of heat and in presence of free alkali. It may be formed, therefore, under damp storage conditions for alkali in the form of calcium hydroxide or ammonia is always present. Cyanamide and dicyanodiamide are both harmful to plants and nitrifying bacteria, and the latter compound, being rather stable, may persist in the soil for a long time. It is therefore important, in using calcium cyanamide as a fertilizer, to avoid conditions which are favourable to the formation of dicyanodiamide—such as uneven distribution and poor incorporation in the soil or application to a soil which is very wet or very dry, or deficient in these catalytic agents which speed up the decomposition of free cyanamide. Undrained peat soils and sandy soils short of organic matter commonly do not give good results with calcium cyanamide. The best results are to be expected on soils in good heart in a warm season with moisture conditions suitable for microbiological activity. Then the toxic effect of cyanamide is reduced and its conversion to nitrate is rapid.

These things must be borne in mind if the full benefit from this fertilizer is to be obtained. Unless it is being used as a top dressing to destroy weeds, calcium cyanamide should be applied a week or longer before sowing the seed, according to the rate of application, and thoroughly mixed with the soil. It can safely be applied in the winter months because loss of nitrogen is bound to be small on account of slow nitrification. Because of its content of free lime it should not be mixed with ammonium salts which would be decomposed with the liberation of ammonia. It may be mixed with potassic fertilizers or with basic slag without trouble, but the mixing of calcium cyanamide with superphosphate has been the subject of much study, for this may bring about the reversion of the superphosphate and the evolution of considerable heat with the production of dicyanodiamide. No loss of readily available phosphate need be feared if the reversion does not go beyond the stage of dicalcium phosphate, and this can be secured by limiting the proportion of calcium cyanamide to about 10 per cent. of the mixture. And if the mixture is kept cool by being spread out in a thin layer, it retains a good friable condition for sowing and the bulk of the nitrogen is converted to urea. It is of interest to observe that small proportions of calcium cyanamide (2 or 3 per cent.) are added to mixed fertilizers in North America partly to neutralize the potential acidity of certain nitrogenous fertilizers and partly to improve the mechanical condition of the mixture. Since "water-soluble P_2O_5" is not recognized in the sale of fertilizers in U.S.A. or Canada considerable latitude exists in the amount of calcium cyanamide used so long as the phosphate remains available.

Calcium Nitrate

The second method of bringing nitrogen into combination —that of effecting its union with oxygen at the temperature of the electric arc—has received considerable attention. It will be remembered that when Sir William Crookes in 1898, in his British Association address, warned the world of the rapidly progressive exhaustion of its supplies of combined nitrogen, it was to the union of nitrogen with oxygen that he looked for the future supply of combined nitrogen for the wheat crop, and he showed experimentally how the two gases would burn together at a very high temperature. Not enough heat, however, is given out by the flame to bring more gas up to the ignition point, hence the flame is only continuous as long as external energy is poured in.

The foundation of the arc process of producing nitrates commercially was due to the Norwegians Birkeland and Eyde in 1902, and the industry was developed rapidly by harnessing the water power at Notodden and Rjukan. Air is passed into an electric furnace in which an electric arc, between two U-shaped copper electrodes, is spread by an electromagnet into a disc of flame about 10 feet in diameter. The temperature of the flame is about 3000°C, the hollow electrodes being kept cool by a current of water.

About 15,000 litres of air are blown in per minute at gentle pressure and the issuing gas containing about 1 per cent. of nitric oxide (NO) is at a temperature of about 1000°C. It is cooled and then passed into two oxidizing chambers, where the combination of the nitric oxide with the oxygen of the uncombined air takes place; then it passes into a series of five granite condensing towers. Down the fourth tower, which is filled with broken quartz, water trickles and picks up enough of the nitrous gases to become 5 per cent. nitric acid at the bottom; this is pumped up and trickles down the third tower, the process being repeated until the liquid leaving the bottom of the first tower contains 30 per cent. of nitric acid. In the fifth and last tower the acid is neutralized by limestone and the resulting mixed solution of calcium nitrite and nitrate is treated with enough of the previously-formed nitric acid to convert it wholly into nitrate, the nitrous fumes evolved being led back into the oxidizing chambers. The product is then concentrated until it solidifies as a material containing about 13 per cent. of nitrogen. The arc process of "fixing" nitrogen is comparatively inefficient, and it has been found to be cheaper to use the power to make ammonia and oxidize it to nitric acid.

Pure anhydrous calcium nitrate, $Ca(NO_3)_2$, contains 17·1 per cent. nitrogen, but it is unfortunately very deliquescent which makes it difficult to store and handle. To overcome this trouble, it is mixed with about 5 per cent.

ammonium nitrate, granulated and then packed in waterproof bags. Although not much used in this country, it is popular in N.W. Europe and is an ideal type of fertilizer since both the basic and acidic parts, the calcium and nitrate, are required by the plant in substantial amounts and no harmful residue is left in the soil.

Ammonium Sulphate (synthetic)

The most effective method of "fixing" atmospheric nitrogen seems to be by combining it with hydrogen to form ammonia, usually called the Haber process. It was developed on a commercial scale just before the first world war and the first factory in Germany was producing about 12,000 tons of combined nitrogen in 1913–14. The war gave a tremendous impetus to the industry because Germany was unable to obtain supplies of Chilean nitrate, of which she had been a large consumer, for the increasing demands for nitrates for explosives. After the war, production continued and expanded on a large scale to meet agricultural requirements and eventually the production of synthetic sulphate of ammonia far surpassed that of the by-product.

This process is less dependent upon cheap electrical energy than upon cheap coal so that plants are generally located in coal areas. The principle of the process lies in the fact that nitrogen and hydrogen combine at high temperatures, and that the reaction is greatly accelerated in presence of a catalyst. Many materials act as catalysts, but some form of iron is apparently the most useful. The reaction is speeded up by increase of temperature but the product is unstable; high pressure is required to improve the stability of the ammonia formed. Hence the efficiency of the plant depends upon a careful adjustment of temperature and pressure and selection of catalytic agent. The process obviously demands great skill from chemists and engineers.

One of the largest plants in the world is at Billingham in this country, and the process adopted there takes place in the following stages:

(*a*) Air and steam are passed over red-hot coke to obtain a mixture of producer gas and water gas, that is to say, a mixture of carbon monoxide (CO), carbon dioxide (CO_2), hydrogen (H_2) and nitrogen (N_2).

(*b*) To remove the CO, the mixture is passed with steam into a saturation tower in presence of an iron oxide catalyst. The reaction which takes place may be represented by the equation:

$$CO + H_2O = CO_2 + H_2$$

The pressure is then raised to 50 atmospheres and the CO_2 is removed by scrubbing with water; the pressure is reduced and the liberated CO_2 is dried and compressed to a solid.

(*c*) The gas mixture, containing 1 volume of nitrogen and 3 volumes of hydrogen, is then compressed to about 250 atmospheres and the last traces of CO in the gaseous mixture are removed by an ammoniacal solution of cuprous formate.

(*d*) The temperature is raised to 500°C in special steel retorts and in presence of a catalyst the two gases combine.

$$N_2 + 3H_2 = 2NH_3 \text{ (ammonia)}$$

(*e*) A strong aqueous solution of the ammonia formed is passed down through a tower, full of aluminium rings, against a current of carbon dioxide gas to obtain ammonium carbonate $2NH_3 + H_2O + CO_2 = (NH_4)_2CO_3$ (ammonium carbonate).

(*f*) This ammonium carbonate solution is mixed with a cream of finely ground anhydrite, ammonium sulphate and calcium carbonate being formed.

$$(NH_4)_2CO_3 + CaSO_4 = \underset{\text{ammonium sulphate}}{(NH_4)_2SO_4} + \underset{\text{calcium carbonate}}{CaCO_3}$$

The mixture is then passed through a continuous industrial filter. The residual sludge of carbonate is washed and dried, and is known as "Billingham Lime." The solution of ammonium sulphate is concentrated in a vacuum until the salt begins to crystallize and the crystals are removed by continuous filtration and passed through a rotary drier and cooler. The type of crystal produced is carefully controlled to meet requirements of storage, transport and so on. The salt is pure white and contains over 21 per cent. of nitrogen.

But the process does not end at this point for further operations are carried out to produce other fertilizers.

(*g*) When calcium phosphate, the chief constituent of rock phosphate, is treated with adequate sulphuric acid, phosphoric acid is produced; if a third of the sulphuric acid is replaced by ammonium sulphate, the reaction may be written as follows:

$$Ca_3(PO_4)_2 + (NH_4)_2SO_4 + 2H_2SO_4 = 2NH_4.H_2PO_4 + 3CaSO_4$$

The calcium sulphate is precipitated as a slurry, and a solution of monoammonium phosphate is obtained. This is the basis of many modern concentrated fertilizers, the pure salt containing 12·2 per cent. nitrogen and 61·7 per cent. "phosphoric acid" (P_2O_5), completely soluble in water. In practice, the reaction is not so simple as the above chemical equation would suggest because the rock phosphate contains many impurities in the form of compounds of aluminium, iron and fluorine which have to be removed. Even here a valuable by-product insecticide is secured in sodium silicofluoride. However, the solution of the ammonium phosphate is concentrated and such substances as ammonium sulphate, potassium chloride (muriate of potash) and calcium sulphate are added, according to the desired composition of the final product, and the suspension is passed on to chilled rolls from which the solid is removed by knives in flakes which are dried and screened to give an uniform granular fertilizer whose percentage of plant nutrients is much higher than can possibly be obtained from a mixture of the so-called "straight" fertilizers.

(*h*) Yet another stage is the production of nitric acid by passing ammonia from stage (*d*) mixed with air over a platinum catalyst

$$\underset{}{NH_3} + \underset{\text{oxygen}}{2O_2} = H_2O + \underset{\text{nitric acid}}{HNO_3}$$

(*i*) The nitric acid may, in turn, be neutralized by more ammonia to form ammonium nitrate

$$HNO_3 + NH_3 = NH_4NO_3$$

Ammonium nitrate is a very deliquescent salt but contains 35 per cent. nitrogen. It is therefore attractive as a highly-concentrated fertilizer, containing half of its nitrogen as the ammonium and half as the nitrate radicle. But there are difficulties in putting it on the market in a form which will reduce its propensity to absorb moisture and its liability to explode.

(*j*) At Billingham a very concentrated solution of the ammonium nitrate is mixed with dried calcium carbonate (*f*) and sprayed from the top of a high tower. The semi-solid mixture is thus cooled and hardened and is then dried, crushed and passed over screens to give the fertilizer called nitrochalk containing about 16 per cent. nitrogen. This greyish product is stable, has good physical condition and is easy to handle. The presence of about 50 per cent. calcium carbonate is an effective method of preventing the ammonium salt from making the soil acid and its general effectiveness has made it a most popular fertilizer for top dressing purposes. In Germany the method adopted to overcome the disadvantages of pure ammonium nitrate was to mix it with ammonium sulphate to give the fertilizer called Leuna Saltpetre containing 26 per cent. nitrogen, of which one-third was in the nitrate form.

The Billingham processes have been described in some detail to illustrate the ingenious cycles of operations adopted in a modern industrial plant, with strict attention to the possible utilization of by-products, leading in this case to the production of an array of materials of outstanding value to agriculture. They also serve to demonstrate the present tendencies to increase the proportion of available plant food in a fertilizer, which reduces transport and labour costs, and to produce fertilizers in granular form, which improves their keeping quality and facilitates their distribution in the field.

Production Costs

It is very difficult to make a fair comparison of the costs of the three fundamental methods of marketing nitrogen—as sodium nitrate, as by-product ammonium sulphate, and as some synthetic compound from atmospheric nitrogen. The first involves factors not common to the other two whilst the second is incidental to the production of the primary products, coke or gas. The third includes three distinct methods of nitrogen fixation—the cyanamide, the arc and the ammonia processes—and the method adopted depends in large measure upon the locality in which the plant is set up. Furthermore, the ammonia process may obtain hydrogen from water-gas, coke-ovens, electrolytic plant or oxidized natural gas. Plant capacity is necessarily limited when electrolytic hydrogen is used because of the size of electrolytic cells available at present. The water-gas ammonia method depends upon large-scale operation and cheap coal and coke for economy. Coke-oven hydrogen plants also require cheap fuel, but may be operated economically in small units and have the lowest capital costs. Cheap power must be available for the arc and cyanamide processes. Various estimates published in 1930 indicated that the ammonia processes were the cheapest and the arc process the most expensive in both capital and operating costs. But so many diverse factors are concerned in such calculations, and so many developments have taken place since 1930, that a discussion of the detailed data would be out of place in this book.

Urea

One other synthetic nitrogenous compound may be mentioned, although its use as a fertilizer has not reached large proportions. This is urea which may be manufactured by heating an aqueous suspension of calcium cyanamide with carbon dioxide and filtering off the calcium carbonate. The solution of cyanamide is then acidified by sulphuric acid and warmed; the excess acid is removed by chalk and the urea crystallized and dried. These reactions are essentially those which have already been described under the decomposition of calcium cyanamide. The interesting feature of this is that urea, a characteristic product of animal metabolism and excreted in the urine, was the first so-called organic compound to be synthesized from inorganic constituents. This synthesis in 1828, followed by many others, proved beyond doubt that the same chemical forces were involved in living as in dead material.

Pure urea contains 46·7 per cent. of nitrogen, the commercial product about 46 per cent. It is readily soluble in water and is rapidly converted into ammonium carbonate and then into nitrate in the soil. It is, therefore, a highly concentrated and valuable nitrogenous fertilizer, having no deleterious secondary effects and suffering only from the fact that it is so concentrated that it presents difficulties in even distribution when used alone.

Waste Products as Nitrogenous Manures

In addition to the definite compounds which have just been described, a very large number of waste products from various industrial or manufacturing processes dealing with material containing nitrogen are employed as nitrogenous manures. For example, all animal tissues contain nitrogen, hence the residues from slaughter-houses, fish-sheds and other processes concerned in the preparation of food, which are not utilizable in other ways, are available for manure. Again, all industries dealing with wool, silk, hair, feathers and skins, give rise to highly nitrogenous waste material. From their origin and mode of preparation it follows that these substances must be of very variable composition: the supply is also apt to be irregular and limited, so that their use tends to be local and confined to particular classes of farmers. Most of the manures of organic origin contain phosphorus as well as nitrogen, but as a matter of convenience some of them may be treated as purely nitrogenous fertilizers, leaving others to be deal with among the compound substances in Chapter IX.

Soot. Of all waste materials, that most generally used is

probably soot; its value, which is due as much to its physical effects upon the soil as to its fertilizing constituents, has been known for the last three centuries at least. It has already been pointed out that coal contains one per cent. or more of nitrogen; in a fire, some of this is evolved as ammonia when the coal is heated, and if it escapes combustion in the higher levels of the fire it is afterwards partially arrested by the particles of carbon constituting soot, which possess an exceptional power of condensing gases upon their surface. In the main, soot is only an impure form of carbon; its fertilizing value is due to the small and variable proportion of ammonia it has thus absorbed from the gases in the chimney. The percentage of nitrogen present may be very low in boiler soot; domestic soot usually contains from 3 to 6 per cent. nitrogen.

Since the nitrogen is present in the form of ammonia, soot as a fertilizer may be regarded as akin to the ammonium salts; its action, however, is profoundly modified by its physical condition. In the first place, the dark colour of soot makes it a very effective absorbent of the sun's rays, so that in sunlight the temperature of land which has been darkened by a sprinkling of soot will rise two or three degrees above that of the same land uncoloured. And as the radiation from such darkened soil at low temperatures is not increased in the same proportion, there is no corresponding loss of heat at night from the sooted land to discount its higher temperature by day. The increased temperature of the soil is particularly valuable in forwarding the growth both of the plant itself and of the bacteria which are rendering available the reserves of plant food in soil. Soot also helps materially to lighten the texture of heavy soils, and on that account is much valued by market gardeners in districts like Evesham, where the land is somewhat clayey and retentive.

Soot is usually sold by the bushel, which weighs about 28 lb., and the lighter the soot is per bushel the more it is valued, because this indicates its purity and freedom from ashes or other admixture. This is probably the best test the grower can apply, for no guarantee can very well be given as to its composition, so small and irregular are the parcels from which any bulk of soot is made up.

Shoddies. Another group of substances which are practically purely nitrogenous manures are the shoddies and kindred products derived from textile industries and other trades dealing with silk, wool, hair, fur, or skin. Properly speaking, shoddy should consist of the short, broken fragments of wool which are rejected in the various processes for preparing woollen fabrics because they are not long enough to make up into yarn, but now the term is applied more generally to any form of waste from silk or wool manufacturing which is no longer profitable to work up for cloth. The material is thus extremely variable in composition; pure wool and silk contain about 16 per cent. of nitrogen, and at one end of the scale of shoddies come materials like carpet waste, cloth clippings, and gun wad waste, which are nearly pure and may contain as much as 14 per cent. of nitrogen. Less valuable, because of the greater admixture of dirt, are wool-combings, flock dust, and other cloth wastes where cotton is also used. These may have 5 to 10 per cent. of nitrogen; while lower still come the manufacturing dust from textile factories, the sweepings of workshops, etc., in which the nitrogen may fall as low as 3 per cent.

Closely allied to such shoddies are hair and fur waste, skin waste, rabbit flick (ears, tail, feet, and other fragments of rabbit skins), feathers, ground hoofs, horn shavings, hoof and horn meal, and leather dust.

All these materials consist in the main of complex nitrogenous bodies called proteins and, being epidermal tissues, must be very stable to fulfil their proper function in the life of the animal. They are quite insoluble in water and not readily attacked by dilute chemical reagents; in the case of leather, the protein has been rendered even more resistant to break down by the action of vegetable tannins or inorganic salts in the tanning process.

Dried Blood.—On the other hand, the nitrogen in dried blood becomes rapidly available to the plant because the protein of blood is necessarily quite different in type and is readily decomposed in the soil. The fresh blood is commonly treated with iron (ferric) salts or by lime and the precipitate dried and screened. It contains from 7 to 14 per cent. nitrogen, but the total production is not large and it is expensive on account of the heavy demand for it.

As a result of the variable composition and character of these animal products it is impossible to generalize about their action as manures, though certain principles may be laid down. Apart from blood they are generally slow and lasting manures, akin in this respect to the more resistant constituents of farmyard manure, but the speed of their action will depend to a very large extent upon the fineness of their division and to the warmth and the amount of cultivation the soil receives. Fine woollen material, like flock dust, rabbit hair, and small feathers, decay with some rapidity in the soil, and give a very considerable return in the season of their application, as may be seen from the following table of results obtained at Rothamsted with a fine flock dust shoddy containing 12·6 per cent. of nitrogen. In Table 8 the results of four years' experiments with different crops are reduced to a common standard, the unmanured plot each year being reckoned as 100, and the effect of the manure is shown for the four successive crops following the application:

TABLE 8.—VALUE OF RESIDUES FROM PREVIOUS APPLICATIONS OF SHODDY. ROTHAMSTED.

	Un-manured.	Shoddy, same year.	Shoddy, previous year.	Shoddy, 2 years before.	Shoddy, 3 years before.
Swedes	100	143·1	...	...	...
Barley	100	166·9	139·9	...	...
Mangolds	100	140·8	136·9	121·9	...
Wheat	100	177·2	147·3	107·5	110·4
Swedes	100	130·7	146·9	126·6	108·1
Mean	100	152	142	119	109

Many of the coarser materials, like rags, hair and skin, may be found in the soil apparently but little changed for a year or two after their application; while such coarse and tough material as crushed hoofs and leather waste must change with extreme slowness, and can be of little service except in such cases as vine borders where the prime cost is not of very great moment but the land has to remain without further manuring for many years. The presence of oil in a sample of shoddy is sometimes regarded as detrimental, since it hinders the access of water and so delays the decomposition of the nitrogenous material. But considering how rapidly many oils and fats are attacked by bacteria, it is doubtful if this objection is valid.

At the present time, though shoddy is used to some extent in general farming in the neighbourhood of cloth-manufacturing districts, and though a certain amount is worked up into compound manures, it is mainly consumed by the hop and fruit growers. Such farmers are dealing with a perennial crop, the quality of which is important; in consequence they prefer a nitrogenous manure which will come into action steadily and continuously throughout the season, rather than an active one which may at any time induce a sudden rush of growth. As the plant continues on the same ground year after year, the residues of slow-acting manures which are not recovered in the first crop accumulate in the soil. Eventually the land becomes stored with manurial residues, which come into action—i.e. decay and nitrify—*pari passu* with the growth of the plant, because both the plant and the soil bacteria are similarly affected by the variations in such factors as warmth and moisture. The result of the continuous and steady feeding of the plant in this fashion is an equable development, which is found to give rise to high quality in the product.

The organic matter present in shoddy is of value in improving the texture and water-retaining power of the soil, and 1 to 2 tons, according to the nitrogen it contains, are regarded as a fair equivalent for 20 tons of farmyard manure, though the latter will supply considerably more non-nitrogenous organic matter. Shoddy is only suitable for arable land, and should preferably be applied in the early winter and ploughed or dug in as soon as possible after it has been spread.

The inevitable irregularity in the composition of shoddy, even in the output from week to week from a single factory, renders its sale on any exact basis a matter of some difficulty. It is, indeed, a very unsatisfactory task to obtain a sample of a few pounds which will properly represent the bulk of a consignment, and the difficulties are renewed in the laboratory when the large sample has to be reduced to a few grammes for analysis. When, therefore, shoddy is bought and sold on a guarantee, a somewhat wide margin of variation must be allowed; a large bulk is, perhaps, best purchased on the basis of a given price per unit of nitrogen, samples being drawn for analysis from each consignment on arrival and an average taken in order to fix the price (Chapter XIII). While nothing but an analysis will afford a definite idea of the quality of a shoddy, some opinion can be formed by tearing a small sample to pieces and trying each portion in a gas or candle flame. Wool, silk, hair, and all nitrogenous materials, frizzle up and burn slowly with an unpleasant smell; cotton, linen, and similar substances of no fertilizing value, burn quickly with a clear flame, since they consist, when pure, of cellulose. Or the mass may be digested by gentle heating with a strong solution of caustic soda or potash, in which the wool and kindred substances will dissolve, leaving untouched the cellulose and dirt. But analysis forms the only real basis for determining the richness of the material, added to which the farmer must exercise his own judgment about its fineness and the possibility of getting it properly distributed throughout the soil.

It would be difficult to enumerate all the bodies which from time to time are applied to the land as nitrogenous manures; leather in the form of dust, turnings and shavings of horn, meat and cheese that have been condemned for food, all find their way from time to time either to the manure manufacturer or to the land.

The only general rule one can apply to such residues is to buy them on their approximate nitrogen content, paying a low unit price because of their slowness of action, and also to take into account the comparative fineness of division and ease of spreading. Even the most resistant material will eventually decay if it is only finely enough divided and thoroughly disseminated throughout the soil.

CHAPTER IV

THE FUNCTION AND COMPARATIVE VALUE OF NITROGENOUS MANURES

Effects of nitrogen—Response to nitrogen, yield and composition, cereals, grass, kale, root crops—Comparison of different nitrogenous fertilizers, cation exchange, soil acidity, soil texture, suitability, cost, experimental results—Recovery of nitrogen.

BEFORE passing on to a comparison of the values of the different nitrogenous manures, it is necessary to consider how far nitrogen exerts on the plant a specific effect that shows itself whenever there is either an excess or deficiency of the constituent in the soil. In the first place, nitrogen is mainly concerned with the vegetative growth of the plant, with the formation of leaf and stem that are the necessary preliminaries to complete development. A deficiency of nitrogen results in a stunted general growth, in which the grain or seed bears a high proportion to the whole weight of the crop; the plant on analysis, however, shows no marked lack of nitrogen as compared with the other constituents. As the amount of available nitrogen is increased, the development of leaf and shoot increases, their green colour deepens, and maturity becomes more and more deferred, so that a crop grown on land over-rich in nitrogen always tends to be late and badly ripened, and to show a profusion of leaf—characters which, in the case of a grain crop, often result in lodging before harvest.

C*

Effects of Nitrogen.—The effect of an application of a soluble nitrogenous fertilizer is very rapid and the deepening in the colour is associated with an increase in the amounts of nitrogen and the pigment carotene—a precursor of vitamin A—in the leaf. The efficiency of the leaf in carbon assimilation is not increased so that the greater growth of the plant is due to the larger area of leaf produced. When the addition of fertilizer is small (about 20 lb. per acre of nitrogen), the increase in leaf area brings about corresponding increases in the formation of carbohydrates and the absorption of nutrients from the soil so that the final composition of the dry matter of the plant may not be much altered. But when larger additions are made the proportion of nitrogen in the leaf rises and, unless its efficiency is increased by proportionate absorption of potassium so as to maintain a suitable ratio of carbohydrate to nitrogen, it becomes soft and susceptible to attack by disease organisms. It is now well established that when increasing dressings of nitrogenous fertilizers are supported by appropriate dressings of potassium compounds, the health of the plant does not suffer and the proportion of carbohydrate in the dry matter is maintained. This is of particular importance in certain root and fruit crops.

One of the most important effects upon plants of an excess of nitrogen is their increased susceptibility to fungoid attacks of all kinds. For example, rust (*Puccinia glumarum*) is much more common upon wheat which has been heavily manured with nitrogen, just as it appears on normally manured wheat whenever the character of the season has been such (heat and moisture together in May) as to induce a specially rapid production of nitrates while the plant was making its growth. In seasons when rust is prevalent, the high nitrogen plots at Rothamsted are always markedly the more rusty, and can easily be picked out by their colour; the grass plots are also marked by their special rusts; and, again, such a characteristic grass fungus as *Epichloe typhina* is generally common enough on the high nitrogen plots but absent from the others. But susceptibility to disease brought about by an excess of nitrogen is perhaps most strikingly seen at Rothamsted on the mangold plots, though the mangold is a plant which, as a rule, suffers but little from fungoid attacks. In September, however, the leaves of the mangolds at Rothamsted that receive an excess of nitrogen begin to be attacked by a leaf spot fungus, *Uromyces betae*, which develops rapidly until on the worst plots all the larger leaves turn brown and present a burnt-up appearance, because the spots of destroyed leaf tissue have become so numerous as to run together. Where the application of nitrogen has been less heavy but is still high, the severity of the attack is diminished, while the fungus is entirely absent from the leaves of the normally manured plots, although they are in close proximity and equally exposed to infection. The association of high nitrogenous manuring with susceptibility to disease may be seen in all plants; it is often manifest in greenhouses where crops are grown in specially rich soil, in which nitrification is very rapid owing to the high temperature prevailing. The dark green aspect of the leaves of such plants is generally evidence of the excessive amounts of nitrogen they are receiving, and it is well known that if any fungoid disease makes its appearance it is very difficult to keep in check; it may destroy the whole crop with great rapidity as, for example, has been the case with the leaf spot fungus *Cercosporium melonis*, which has in some years proved so destructive to cucumbers grown under glass.

Various attempts have been made to find an explanation for the cause of this association of high nitrogenous manuring with susceptibility to disease. In the first place, certain physical differences can be traced in the tissues of the plants; just as high nitrogen results in a weakness of straw in cereals, due to a long-jointed soft stem, so the cuticle of the leaf and the cell walls of the leaf tissue are measurably thinner when the plant has been grown with an excess of nitrogen. The cause may, however, be found in some alteration in the composition of the cell sap, which renders it a better medium for the growth of the fungus in question. It has been found, for example, that spores of the *Uromyces betae* will grow freely upon a bruised surface of the mangold leaves grown with excess of nitrogen, but make no headway when sown upon a similarly bruised surface of the leaf of a normally manured plant.

The softness of tissue which is induced by large applications of nitrogenous manure is recognized in other ways; for example, cabbages and similar vegetables grown rapidly with excess nitrogen are preferable for immediate consumption because of their tenderness, but in the market they bear a bad reputation, because the same softness of tissue leads to rapid wilting and a faded appearance when the vegetables have been cut for some time and have experienced the usual amount of rough handling in transit.

On most of our cultivated soils, when the cropping is continued and manure withheld to a point when there begins to be a serious falling off in the yield through lack of plant food, it is the want of available nitrogen rather than of phosphorus and potassium which determines the yield; in other words, the soil is much more rapidly exhausted of its available nitrogen than of its available phosphate and potassium. Thus, while each of these three constituents of plant food is equally indispensable to the plant, good crops can often be grown by the aid of a nitrogenous manure alone. The special value of nitrogen in this connection is well seen in the Rothamsted experiments; on the wheat field, for example, we may compare the yield of the unmanured plot with that receiving nitrogen alone and minerals alone, and again that which receives nitrogen and phosphate against that which receives nitrogen, phosphate and potassium.

From Table 9 it will be seen that Plot 5, which is nitrogen starved but which receives an excess of all the other elements of nutrition, yields only 1·9 bushels more grain than the unmanured plot; whereas Plot 10, which receives an excess of nitrogen but has had to rely solely upon the original reserves of minerals in the soil, has produced on the average 7·6 bushels of corn more than the unmanured plot. The minerals increased the yield by 14·7 per cent., but nitrogen by

TABLE 9.—AVERAGE YIELD OF WHEAT. BROADBALK, ROTHAMSTED. 56 YEARS (1852–1907).

Plot.		Grain.	Straw.
		Bushels.	Cwt.
3	Unmanured	12·9	10·5
5	Mineral Manures only, no Nitrogen .	14·8	12·3
10	Nitrogen only, no Minerals . .	20·5	18·7
11	„ and Phosphates. . .	23·7	22·8
13	„ Phosphates, and Potash .	31·6	31·9

59 per cent., and these differences would have been much more pronounced had they been calculated on the results of the first year or two of the experiments only, instead of over a period so long that the mineral reserves of the soil are also highly exhausted. It is this greater relative deficiency of available nitrogen than of available potassium or phosphate in the soil which makes the nitrogenous manures so important in practice.

Response to Nitrogen

In assessing the value of any fertilizer, and this applies particularly to nitrogenous fertilizers, it is important to keep in mind the influence of climate before and during the growing season. In a mild moist winter, such as commonly obtains in this country for example, a large quantity of soluble nitrogen compounds is leached to the drains and this is likely to reduce the subsequent yield of autumn-sown wheat. The winter rain not only washes out nitrates in the soil but packs the soil and reduces aeration and therefore biological activity in the early spring months. Consequently, the wheat is specially responsive to a dressing of nitrogen in the spring and the wetter the winter months the greater the response. Wheat does not experience the same difficulty in obtaining sufficient phosphate or potassium for it develops an extensive root system and has a long period of growth. In comparison, spring-sown cereals have a shorter growing season and a more restricted root system, and their chief growth is made later in the season on a soil which has been well aerated by cultivation. Nitrification in the soil, therefore, plays a greater part in supplying available nitrogen, but the search for other nutrients is more limited and barley and oats often show a greater response to phosphatic and potassic fertilizers than wheat. A similar contrast is to be found with mangolds and turnips. Mangolds are sown earlier and are deeper rooting and, generally speaking, respond more to nitrogenous fertilizer and less to phosphates than do turnips. In fact, swedes show a greater response to phosphate than any other crop.

In addition to the variations due to climate and crop, there are to be considered the factors of soil fertility and the interaction of one fertilizer with another. Naturally, one does not expect the responses to fertilizer to be as marked on a rich soil as they are on poor worn-out land or on land which is in bad physical condition, for many interdependent factors are involved in the satisfactory development of a plant, and any one of them may be a limiting one. In this respect, the response to one fertilizer may be governed by the amount of another. As might be expected, the effect of nitrogen is usually increased when phosphatic or potassic fertilizers are also used because the increase in vegetative growth means an increase of all the processes concerned in carbon assimilation by the leaves and absorption by the roots. These interactions vary, of course, from crop to crop and from field to field and depend upon the extent to which the soil is capable of providing the nutrients required by the plant. The popularity of certain compound manures—adjusted in composition for different crops—is simply a convenient insurance against the soil not containing enough of one or more nutrients. It has sometimes been described as a case of "adding a little of everything in the hope that something will work," and it is often not the most economical method. The fuller use of the results of manurial experiments and soil analysis will, in future, improve the efficiency of manuring.

Yield.—Meantime, for average soils in northern Europe, we

have a very good picture of the fertilizer requirements of arable crops in the comprehensive study by Crowther and Yates in 1941 of experimental results published since 1900. With due regard to the facts that the results are averages covering many soil types and do not take into consideration residual effects, the responses to the main kinds of fertilizer by different crops were found to be remarkably consistent. With respect to nitrogenous fertilizers, all the responses were reduced to those given with 28 lb. nitrogen per acre, equivalent to 1·2 cwt. ammonium sulphate. For nearly 7,000 experiments this treatment showed average gains of 3·8, 4·1 and 3·6 cwt. grain per acre respectively for wheat, barley and oats. For Great Britain alone, the responses were rather less, being 3·4, 3·7 and 3·4 respectively. These responses are of the order of 20 per cent. of the crop yields and are essentially the same for the three cereals. Only cereal experiments without dung were considered.

In the case of the root crops it was necessary to take into consideration the use of dung. Nearly 13,000 experiments with dung were summarized and the average responses in tons per acre to 28 lb. nitrogen were swedes 2·1, mangolds 2·8, sugar beet 1·3, and potatoes 0·85. For 900 experiments without dung in Great Britain, the corresponding figures were 2·2, 3·1, 0·88 and 1·07, and for 700 experiments with dung they were 2·3, 2·6, 0·92 and 0·86. These responses are of the order of 10 or 12 per cent., and it is of particular importance to observe how little they are influenced by the presence or absence of dung.

The response to a fertilizer is not proportional to the amount applied, except for very small applications, but falls off in conformity with the law of diminishing returns. For example, Crowther and Yates have shown that in the case of potatoes the average response is 0·8 ton to 22 lb. nitrogen, 1·4 ton to 44 lb. and then rapidly decreases so that it is only 2 tons to 112 lb. nitrogen. They have prepared curves (p. 248) for all the main crops, from which it is possible to estimate the most profitable dressings of nitrogen. These are necessarily related to the current cost of fertilizer and the value of the crop but (for 1954 prices) they are generally about 50 per cent. larger than the dressings which are actually used in practice, and this excludes cases where fertilizer is not used at all. From these figures it is quite obvious that crop yields could be very considerably increased by the more liberal use of nitrogenous fertilizers. In the case of the cereals, fear of lodging is, of course, one reason for the difference between the dressings actually given and those which produce larger crops for, as the amount of nitrogen applied increases, the ratio of straw to grain also increases. In other words, the production of straw increases more than the production of grain. To quote one example from the Broadbalk wheat field, as the dressing of nitrogen was increased from nothing to 172 lb. per acre, the ratio of grain to straw decreased steadily (average from 1852 to 1864) from 0·62 to 0·48. But the effect of nitrogen on the yield of straw is largely determined by the time of application; in fact, a late dressing may show up only in the composition of the grain and have little influence on the amount of straw produced, or on the yield of grain. Hence, generally speaking, heavier dressings of nitrogen applied early in spring increase the total yield but tend to give a greater proportion of straw to grain with consequent risk of lodging, whereas late dressings do not make much difference in the proportion of straw to grain but alter the composition of the grain.

Cereals.—The composition of the cereals has received careful study in many parts of the world. For example, in North America it was established that, by delaying the application of nitrogenous manure to wheat until heading time, a high protein flour of superior baking quality was obtained. In this country, however, although it has been demonstrated that a late application of nitrogen increases the percentage of protein in the grain, the effects on the baking quality of the flour have not been sufficiently promising to attract attention.

In the case of barley, on the other hand, the investigations have led to the most valuable conclusions in the production of grain for malting purposes. A nitrogenous fertilizer increases the yield of grain and the effect seems to be greatest when the dressing is applied at seedtime. Like other cereal grains, the composition of barley shows great variations due to variety, season and environment but a moderate dressing of nitrogen (about 1 cwt. ammonium sulphate) may have little effect on the protein content provided the crop is well supplied with phosphate and potassium. With dressings of 2 cwt. or more, the percentage of nitrogen in the grain increases rapidly and the glutelin nitrogen is a fairly constant fraction (36 per cent. in Plumage Archer) of the total nitrogen for any variety. The value of the grain for malting purposes is inversely related to its nitrogen content; that is to say, the nitrogen content should be fairly low and the starch content high. To secure such grain, the barley must be grown under favourable conditions and a moderate dressing of nitrogen can be applied at seedtime to obtain a satisfactory yield without danger of reducing the quality.

The valuation of oat grain does not seem to be closely related to its nitrogen content which can be increased by nitrogenous manuring with a distinct gain in food value. In an investigation in the east of Scotland, dressings of nitrogen varying from 17 to 60 lb. per acre, as sodium nitrate or ammonium sulphate, were applied at different times from seeding until the middle of June—as late as it was practicable without permanent injury to the crop. During the six years in which the experiments were carried out, the percentage of nitrogen in the grain was invariably increased by delaying the application of nitrogen. The results showed that the response of different varieties on different soils was both general and significant, the average increase being 6·5 per cent. There was no regular effect on bushel weight or on the dry matter content of the grain, and all the available evidence showed that, although the quantity of straw might be reduced by delaying the dressing, the yield of grain was not much altered. Many of the plots were laid down in "late" districts to test the time of ripening and danger of lodging. Ripening was delayed somewhat by the late dressing but, where lodging took place, it was usually on those plots which received an early dressing and which usually carried a longer straw. The question of delaying the dressing of nitrogen, which is commonly given to oats at time of sowing, until growth is well advanced, therefore seems to be worth further consideration as a means of increasing the protein in this home-produced foodstuff.

Grass.—Striking results are commonly obtained by the application of nitrogenous fertilizers to grass and the responses are extremely rapid. For example, the protein content of hay may be increased 30 per cent. or more by the application of 28 lb. nitrogen about 3 weeks before cutting, and there is a very close parallelism between protein and carotene. But the effects on grazing land are somewhat complicated by the fact that although there is an increase in the proportion of leaf to stem and a longer period of growth, there is a reduction in the protein-rich clover in the herbage with the result that the benefits of the nitrogen may be cancelled out. In addition, an increase in the protein of the grass is generally accompanied by a decrease in its content of calcium and phosphorus which are so important to the growing or milking animal. Consequently, the liberal use of nitrogen must be accompanied by careful attention to liming and to phosphatic and potassic manuring, and to good grazing management. There is no doubt that young grass, especially in the spring, is a particularly valuable foodstuff by virtue of its large proportion of high-class protein; as a result of intensive investigations at many centres in this country, the influences of manuring and grazing on preserving high quality pasture are now well understood.

Kale.—This is another crop which responds well to nitrogen; even to such extremely large amounts as 100 or even 200 lb. per acre, without sign of diminishing returns from successive increases in the size of dressing. Increased yields of up to nearly 100 per cent. have been recorded under experimental conditions and, although the proportion of dry matter tends to fall somewhat in these circumstances, there is no doubt about the value of this crop for converting inorganic nitrogen into leafy material rich in protein for stock feeding. Crops of more than 25 tons per acre of marrow-stem kale have been common, yielding about 10 or 11 cwt. per acre of crude protein.

Root Crops and Potatoes.—Nitrogenous fertilizers increase the leaf area as with other crops, but here we are concerned primarily with the efficiency of the leaf in building up the store of carbohydrates in the root or tuber. The distribution of dry matter between leaf and storage organ is, therefore, of greater importance than the mere production of much leaf; in general, it may be said that, in the response to nitrogen, the ratio of root to leaf is better the longer the period of growth after full development of the leaf. For that reason nitrogenous fertilizer is best applied early in the season, usually at time of sowing, and has a greater effect on the yield of mangolds than of sugar beet or turnip roots. The average responses of these crops to 28 lb. nitrogen have already been mentioned but, provided proper attention is paid to the other plant foods, much larger dressings of from 60 to 90 lb. nitrogen may be profitably applied except in the case of swedes where the larger dressings are inclined to show mainly in a larger growth of leaf which is less efficient and more susceptible to damage. Average responses, in absence of dung, are of the following order: mangolds, 5 tons to 66 lb. N; sugar beet, 1·8 tons to 82 lb. N; swedes, 2·2 tons to 28 lb. N; potatoes, 1·9 tons to 95 lb. N.

The effects are not of course to be measured simply by these total yields of roots or tubers. In the case of mangolds, excess nitrogen depresses the proportions of dry matter and sugar in the roots and increases the percentage of nitrogen in the dry matter. But, in normal agricultural practice, this scarcely affects the large responses due to nitrogenous fertilizer. With swedes, also, the reduction in dry matter percentage is relatively small compared with the percentage increase in yield when the fertilizer is judiciously used, and with potatoes the percentage of dry matter in the tubers is not much altered. The yield of sugar is the important criterion when sugar beet is considered and, although dressings of nitrogen of about 80 lb. per acre depress the percentage of sugar in the root by about 0·4, the increase in yield of washed beet is sufficient to account for an actual increase in total sugar of about 3 to 4 cwt. in an average season. In addition, the increased yield of tops, amounting to 2·5 to 3 tons, available as cattle food, should be taken into account in estimating the return obtained from the fertilizer.

Comparison of Different Nitrogenous Fertilizers

Cation Exchange.—Before proceeding to discuss the relative values of the various fertilizers, it is necessary to consider their fate in the soil. When any salt solution is brought into contact with the very fine colloidal particles of clay or humus in the soil, a reaction called cation exchange takes place. The colloidal clay particle may be conveniently regarded as an alumino-silicate on the surface of which are absorbed calcium, magnesium, potassium and sodium; there is also hydrogen in a humid climate unless the soil contains excess calcium carbonate as limestone or chalk. These are referred to as the exchangeable cations, for in presence of a solution containing a cation, such as sodium in sodium nitrate or potassium in potassium chloride, there is an exchange of cations between the clay (and humus) and the salt solution. The solution now contains some calcium and magnesium in addition to potassium or sodium, while the clay contains a larger proportion of sodium or potassium among its exchangeable cations than it did before the reaction. The exchange takes place very rapidly, and the extent to which it goes depends upon the final concentration of the various ions in solution; but the negative or anion part of the salt—the nitrate or chloride in this case—remains unaltered. Hence, when a fertilizer salt like sodium nitrate is added to a soil and goes into solution, the composition of the solution is soon changed by the substitution of calcium and other cations for some of the sodium which is absorbed by the soil. In other words, the addition of a sodium salt brings within reach of the plant several other soluble ions. Any ions which are not absorbed by the growing plant go to change the composition of the drainage water.

Soil Acidity.—As a result of biological activity in the soil considerable amounts of carbon dioxide are liberated which dissolve in water to give carbonic acid. The hydrogen of the carbonic acid behaves like the metallic cations and displaces calcium, magnesium and so on from the soil. In a normal soil, calcium is the predominant exchangeable cation and considerable amounts of calcium are lost to the drains in the form of soluble salts, particularly bicarbonate, by this exchange process. The loss varies greatly from soil to soil, but usually amounts to the equivalent of 1 to 5 cwt. lime per acre per annum. Hydrogen takes the place of the calcium in the soil and is responsible for the condition known as soil acidity. To prevent the acidity from becoming serious, it is essential to reverse this normal reaction by adding, from time to time, a calcium compound—usually in the form of lime (which is calcium oxide) or limestone or chalk (which are both calcium carbonate). The need for some kind of liming material at regular intervals is inevitable unless the soil happens to contain free calcium carbonate.

When ammonium sulphate is added to the soil, the ammonium ion behaves in the first instance exactly like sodium or potassium; some of it displaces calcium so that the soil solution now contains calcium sulphate and the soil particle some absorbed ammonium. But the ammonium is attacked by micro-organisms in the nitrification process; it is oxidized to nitric acid which reacts with insoluble calcium compounds to form calcium nitrate; the nitrate anion is not absorbed by the soil but remains in solution and is either taken up by the plant or is leached to the drains. The drainage therefore contains very little ammonium salt but is enriched in other salts—particularly calcium bicarbonate, sulphate and nitrate—while the soil particle becomes acid through the ultimate displacement of calcium by hydrogen. Theoretically, these reactions may be illustrated in the following manner, where X represents the colloidal particle on which the cations are absorbed:

(1) $(NH_4)_2SO_4 + CaX \xrightarrow{\text{cation exchange}} (NH_4)_2X + CaSO_4$

(2) $(NH_4)_2X \xrightarrow{\text{nitrification}} 2HNO_3 + H_2X$

(3) $2HNO_3 + CaX \xrightarrow{\text{neutralization}} H_2X + Ca(NO_3)_2$

the net result being that 2 nitrogens in the sulphate displace 2 calciums from the soil and leave 4 hydrogens which in turn will require 2 calciums to bring the soil back tc normal.

(4) $2H_2X + \underset{\text{limestone}}{2CaCO_3} \longrightarrow 2CaX + \underset{\text{carbonic acid.}}{2H_2CO_3}$

This means that 132 parts of ammonium sulphate would account for 80 parts of calcium or for every 28 lb. nitrogen added in the form of ammonium sulphate 200 lb. of limestone or 112 lb. burnt lime would be required to make good the loss of calcium occasioned by these reactions. But this implies certain assumptions with respect to the condition of the soil and the absorption of nitrogen by the plant, and the actual requirement of calcium carbonate is much less. Various figures have been found in practice and they generally lie between 100 and 120 parts calcium carbonate for every 100 parts of ammonium sulphate.

When we consider monoammonium phosphate a complication is introduced by the fact that the calcium liberated by exchange forms an insoluble calcium phosphate and is not washed out of the soil. The loss of calcium is therefore very much less and similar to that due to the addition of the same amount of nitrogen in ammonium nitrate (where half the total nitrogen is in the form of nitrate) or in urea, which is converted in the soil to nitrate through ammonium carbonate. Nitrochalk consists of ammonium nitrate mixed with an equal amount of calcium carbonate which is more than sufficient to balance the potential loss of calcium by the ammonium ion, so that the exchangeable calcium of the soil is actually increased. Essentially the same result follows the use of calcium nitrate because the nitrate can be directly absorbed by the plant while any calcium ions which are left in solution are free to reduce soil acidity. This effect is even more pronounced with calcium cyanamide. This compound undergoes a decomposition in the soil, leading eventually to ammonium carbonate and then to nitrate, but it contains calcium equivalent in amount to the nitrogen and, furthermore, the commercial material used as fertilizer contains about 20 per cent. of lime.

Soil Texture.—In all these cases we are concerned principally with the loss or gain of exchangeable calcium in the soil colloids. When sodium nitrate is considered the position is somewhat different. The nitrate ion is absorbed by the plant more rapidly than the sodium ion, leaving sodium to take part in a cation exchange or to form sodium carbonate with carbonic acid. In normal practice, the amount of sodium nitrate added as a fertilizer is not large—usually 1 or 2 cwt. per acre—and the sodium ion has a relatively small displacing power. The actual cation exchange is, therefore scarcely measurable, but the sodium ion can neutralize acid ions in the soil solution and so reduce the development of soil acidity and the need for lime to meet the normal losses of calcium from the soil. When sodium nitrate is applied to the soil in excessive amounts then the proportion of absorbed sodium in the clay increases and may reach a point at which the particles are deflocculated; that is to say, the normal favourable crumb structure of the soil, which is the result of the individual small particles forming aggregates and which is maintained by the large proportion of exchangeable calcium in the colloidal particles, breaks down. The permeability of the soil is reduced by the dispersion of the colloid particles and, in wet conditions, the land may poach. It is extremely doubtful, however, whether this takes place to any extent in normal farm practice, and the effect is to be expected only on clay soils.

Extreme Cases.—It is important to guard against exaggerated statements which are frequently made about the acidification of the soil by ammonium sulphate, or about the deleterious effect of sodium nitrate on soil texture. There should be no question of these tendencies becoming serious where good husbandry is practised. The long-continued experiments at Rothamsted and Woburn demonstrate effectively what may happen in extreme cases. On the permanent grass plots at Rothamsted, where the same manuring has been maintained on the same plots since about 1860, the repeated application of 43 lb. nitrogen as ammonium sulphate increased the acidity to a marked extent whereas sodium nitrate reduced it. The increased acidity completely changed the vegetation.

The reason is that under these conditions biological activity becomes depressed, the humus becomes acid in character and tends to accumulate on the surface due to the absence of earth worms, and only those species which do not depend upon nitrogen as nitrate can survive. Leguminous species disappear and narrow-leaved grasses predominate. The vegetation in fact becomes admirably suited to meet the requirements of lawns and golf greens; indeed, acidification of the soil by means of ammonium sulphate is one method adopted in the management of golf courses to produce a closely-knit sward of fine grass free from clover and broad-leaved species.

At Woburn, the effect of annual applications of ammonium sulphate became apparent in about 20 years. Thereafter barley failed; but on the plot which was adequately limed, ammonium sulphate continued to produce a yield of barley comparable to that obtained with sodium nitrate. It is only when the soil contains no free calcium carbonate that the loss of exchangeable calcium, with increase in acidity, caused by ammonium salts, requires close attention. Ammonium sulphate is then responsible for increasing the normal loss of calcium from the soil, but the amount of lime required, in addition to that which should be applied to compensate for normal losses, is not large—little more than half the weight of ammonium sulphate employed.

Suitability.—In selecting one of the various nitrogenous fertilizers, it is important to bear in mind the suitability of the fertilizer for the crop and soil in question, the rate at which the nitrogen becomes available, the possible effect of a change in soil acidity, the cost of the nitrogen and the cost of preventing or correcting any change in the lime status of the soil. In comparing the value of the different fertilizers as a source of nitrogen, due regard should be given to the possible effects of the acidity factor and of the presence of other elements besides nitrogen, which might influence the crop in question.

A few examples may be mentioned. It has already been pointed out (Chap. III) that the best results are not to be expected with calcium cyanamide unless the soil conditions are suitable to promote the speedy decomposition of that compound to the form in which it is useful to the plant; the soil should be well drained, contain a moderate amount of colloidal material and encourage biological activity. Chilean nitrate contains a large amount of sodium which is a valuable nutrient in the case of mangolds and sugar beet; it also contains a little potassium, and traces of a great many other elements, some of which are essential in plant growth. Then most cultivated crops absorb nitrate readily so that the response to the addition of nitrates is very rapid. But excess nitrate is easily leached from the soil; it does not show to best advantage, therefore, when the rainfall is high and may be washed out more quickly than the plant can absorb it.

Many plants can make use of nitrogen in the form of ammonium ions—in fact, some plants must absorb the bulk of their nitrogen in that form when the soil conditions are such as to reduce or preclude nitrification. But although some plants may derive immediate benefit from ammonium salts added to the soil, the rapid response in most cases comes mainly after nitrification has taken place. As already stated, ammonium salts are not washed out of the soil; a loss of nitrogen to the drainage water occurs only if nitrate is formed in excess of what the plant can use. Ammonium salts are therefore to be regarded as more suitable than nitrates in areas of heavy rainfall. Furthermore, the sulphate ion in ammonium sulphate is valuable as a plant nutrient, especially in the case of leguminous crops and members of the Brassica family.

The organic manures are obviously a special case in which the speed of production of available nitrogen is, with certain outstanding exceptions such as dried blood, relatively slow and governed by the nature of the material and its fineness of division. The slow and steady production of nitrate in the soil as the plant is developing is a most desirable feature of these substances; but, except in particularly favourable circumstances, the amount of available nitrogen is usually inadequate at certain periods of rapid growth and should be supplemented by the addition of easily soluble compounds. When easily available nitrogenous fertilizers are being used in relatively large quantities, to maintain rapid growth in vegetable crops, for example, it is obviously desirable to make periodic applications of small dressings rather than a single large one.

Cost.—The price of these various fertilizers and manures has fluctuated considerably in recent years and the only practicable method of comparison is on cost per unit of nitrogen, which means cost of one per cent. of nitrogen in a ton of fertilizer. For example, if a fertilizer containing 20 per cent. N costs £20 or 400 shillings per ton, the cost of one unit of nitrogen is 20 shillings. For some years, ammonium sulphate has been the cheapest form of nitrogen and the grower has had to decide whether the greater price for nitrate-nitrogen was justified by its advantages in his particular case. The nitrogen in organic manures has been the most expensive largely because of the large demands for them. In the early spring of 1955, for example, the nitrogen in dried blood cost about 92 shillings per unit, that in sodium nitrate rather less than 33 shillings, that in ammonium sulphate rather more than 17 shillings. When organic manures are used in substantial quantities of the order to 2 or 3 tons of dry matter per acre, they, of course, make a valuable contribution to the amount of colloidal humus in the soil and so improve both the chemical and physical properties of the soil—an effect which is not to be expected from the inorganic fertilizers. They also possess a residual value in the form of complex nitrogen compounds which have not decomposed and liberated their nitrogen in the first year after application. This is also in contrast to the soluble fertilizers whose nitrogen is practically all used up in the first year. These special advantages of organic manures are, however, barely measurable when the dressings are small, for the addition of even 1 ton of dry organic matter per acre amounts to less than 0·1 per cent. of the soil to plough depth, something very much less than soil sampling errors. Some sense of proportion is therefore called for in computing the value of organic manures. The prices which are frequently paid for them seem to suggest that the purchasers have a most sanguine faith in their efficacy.

From the above discussion it will be clear that it is not possible to say which is the *best* nitrogenous fertilizer. A great variety of factors will play a part in any comparison which may be made. Some results from the long-term experiments at Rothamsted are of great assistance in establishing certain principles.

Experimental Results.—Several of the organic compounds of nitrogen, such as those contained in Peruvian guano, rape cake, and dried blood, are almost as active sources of nitrogen as ammonium salts, especially when used continuously, so that the residues left in any one year are available for succeeding crops. For example, the barley plots receive equal weights of nitrogen as nitrate of soda, sulphate of ammonia, and rape cake; and as the following table shows, the returns from the rape cake are but little below those from the other two manures.

TABLE 10.—NITROGENOUS MANURES WITH MINERALS. AVERAGE YIELD OF BARLEY (1852–1901). ROTHAMSTED.

Plot.	Manuring.	Grain.	Straw.
		Bushels.	Cwt.
4A	Ammonium Salts = 43 lb. N. . .	42·1	25·0
4N	Nitrate of Soda = 43 lb. N. . .	43·6	27·4
4C	Rape Cake = 43 lb. N.	41·0	24·5

To examine the question of residual values, the scheme of experiment is to take four plots for each manure; one receives the manure in any particular year, while the others remain unmanured except for the residues that may remain from a similar application that had been made one, two, and three years previously, a fifth plot continuously unmanured being employed as a check. The experiments have not been in progress long enough to enable exact results to be obtained, especially as regards the residues remaining in the second or third year after the application, but the following figures show the kind of return which may be anticipated. In order to eliminate the effect of season and crop, the increase given by a residue is always compared with the increase brought about by a fresh application of the same manure, which increase is reckoned as 100.

TABLE 11.—INCREASED YIELD DUE TO RESIDUES OF NITROGENOUS MANURES COMPARED WITH INCREASE PRODUCED IN FIRST YEAR. ROTHAMSTED.

	Year of Application.	Second Year.	Third Year.
Dung	100	46	37
Wool Waste . .	100	79	38
Peruvian Guano . .	100	12	12
Rape Cake . . .	100	9	2

Thus rape cake, a manure which we have seen to be comparatively active, leaves behind for the following year a very small residue, having only 9 per cent. of the effect of a fresh application of manure; whereas a wool shoddy increases the crop in the second year by as much as 79 per cent. of the increase produced by a fresh application of the same manure, and even after two crops have been removed the residue is still one-third as effective as a fresh application.

The potent effect of organic manures in promoting a good tilth is very clearly shown by the Rothamsted experiments upon mangolds, where the nitrogenous manures are nitrate of soda, sulphate of ammonia, and rape cake respectively. In a good season the nitrate of soda is the most effective manure; but taking an average over the whole period, rape cake shows a great superiority, simply because of the difficulty of getting a full plant upon the other plots. Though all the plots are cultivated in the same way and at the same time, the condition of the soil has become so bad where purely inorganic manures have been used, that only in favourable seasons is what a farmer would call a good plant obtained on the nitrate and the ammonia plots, whereas the rape cake plot starts regularly enough. On three occasions the plant has completely failed on the ammonia and nitrate plots. Even in the other years there are great deficiencies, as shown by the average number of plants counted on each plot, which is set out in Table 12.

TABLE 12.—EFFECT OF MANURES UPON THE NUMBER OF ROOTS. ROTHAMSTED MANGOLDS, 1876–1902.

Plot.	Manures.	Average Crop per acre.	Average Number of Roots per acre.
		Tons.	Number.
4C	Complete Minerals with Rape Cake . . .	21·3	17,474
4A	Complete Minerals with Ammonium Salts . .	14·9	14,802
4N	Complete Minerals with Nitrate of Soda . .	18·0	14,130

In ordinary farming, the effect upon the soil is never likely to become so pronounced as in these experiments at Rothamsted, but without a doubt a considerable element in the extra value which the farmer sets on organic nitrogen must be put down to its improvement of the texture of the soil, a factor the farmer rightly regards as of the first importance.

With respect to comparisons between the inorganic fertilizers, probably the greatest number of experiments have been devoted to the relative effects of sodium nitrate and ammonium sulphate. It has been recognized for some time that the nitrate is generally better for mangolds and sugar beet and barley, whilst the ammonium salt is better for potatoes and oats. It has always been difficult, however, to base a generalization on the data from different experiments unrelated to each other, often carried out for one season and one crop only, and seldom incorporating more than one rate of dressing. It is generally desirable in fertilizer experiments to obtain some idea of the growth curves by examining the effects of two or more rates of treatment. It is then possible to make a comparison of different materials even although they have not been used at exactly the same rate.

Crowther has recently made a summary of results obtained in over 800 field experiments in North-West Europe, in which ammonium sulphate was compared with an equivalent amount of nitrogen in Chilean nitrate. He made use of the response curves, relating increase in yield to increase in amounts of nitrogen applied, in order to reduce the results to standard responses to the two fertilizers. As far as possible, the data were taken from experiments in which secondary effects would not be pronounced, and the following average results were obtained. The figures show the response to nitrate when that to ammonium sulphate is taken as 100.

Mangolds	147
Sugar Beet	129
Swedes, Rye, Wheat, Barley	118–124
Oats	92
Potatoes	79

These figures suggest that there is little doubt about the superiority of the nitrate for mangolds and of the ammonium salt for potatoes. The question of cost must be considered in the case of the other crops. In recent years, the unit of nitrogen has cost about twice as much in sodium nitrate as in ammonium sulphate (Chapter XIII). This means that, for the same expenditure, 0·25 cwt. nitrogen in the nitrate could be replaced by 0·50 cwt. nitrogen in the ammonium salt. But the response to fertilizer does not increase in proportion to the amount of fertilizer used. In fact, the response curve in the case of ammonium sulphate shows that 0·50 cwt. nitrogen gives only a 50 per cent. greater increase in crop yield than 0.25 cwt. nitrogen. Hence, when the two fertilizers are compared on an equal nitrogen basis (say 0·25 cwt.), the sodium nitrate will only be the more profitable if it shows a response at least 50 per cent. greater than the ammonium sulphate. At current prices, therefore, it will be obvious that ammonium sulphate is actually the cheaper of the two fertilizers for swedes, wheat, barley and rye; sodium nitrate and ammonium sulphate are equally profitable for mangolds; ammonium sulphate is undoubtedly the cheaper for oats and potatoes. These conclusions might have to be modified with a change in current prices of the fertilizers and, of course, they ignore the special characteristics of each, previously discussed. For example, the use of about 150 lb. of these fertilizers would increase the farmer's lime bill by three or four shillings in the case of ammonium sulphate and reduce it by about a shilling in the case of sodium nitrate. At present these results constitute the best available general picture of the experimental evidence for the climate and soils of N.W. Europe.

There have also been investigations on the relative responses to the other nitrogenous fertilizers. Calcium cyanamide and urea seem to be rather less effective than ammonium sulphate, for equal applications of nitrogen, whilst ammonium chloride (muriate of ammonia) gave promise, before it was taken off the market, of being better for malting barley but not so satisfactory for potatoes. Nitrochalk, which contains equal amounts of nitrogen in ammonium and nitrate forms and also calcium carbonate in sufficient quantity to reduce the acidification of the soil, has generally been found to be more effective than ammonium sulphate, and is becoming increasingly popular particularly for top dressing purposes. These observed effects may, of course, be influenced by the use of farmyard manure, and so far there is a serious lack of experimental evidence on this question.

Recovery of Nitrogen

In many investigations, the composition of the crop as well as its total yield has been determined in order to compute the recovery of nitrogen added as fertilizer. The recovery has been found to vary, according to the utilization of the fertilizer by the crop in question, from about 20 to 90 per cent. One pound of nitrogen produces on the average an increase of dry matter amounting to about 13 lb. in cereal grain and 24 lb. in straw, corresponding to approximately ⅓ lb. nitrogen recovered in the crop. The recovery in the swede crop tends to be low, but mangolds and sugar beet may show very high recoveries when sodium nitrate is the source of nitrogen. As a fair estimate for average conditions, it may be assumed that about half the nitrogen added is returned in the crop whilst the remainder of the nitrogen added is partly held by plant roots and soil organisms and partly lost in the drainage water.

There is a common criticism that nitrogenous fertilizers exhaust the land. Since they produce larger crops which have necessarily absorbed from the soil more calcium, phosphorus, potassium and so on than smaller crops, the criticism is justified. But the grower should not expect to continue indefinitely to secure larger crops without returning these soil constituents which are removed. The fertility of the soil can be maintained only if the nitrogenous fertilizers are supported by fertilizers or manures containing the other plant foods.

And it should not be overlooked that the larger the crops the larger the plant residues left in the soil; and, if the crops are used to feed stock, the larger the amount of animal excreta available for return to the soil.

CHAPTER V

PHOSPHATIC MANURES

The phosphates of calcium—Bone manures—Guanos—Mineral phosphates, origin, production—Superphosphate, world production, manufacture, composition, world consumption, reactions—Concentrated superphosphate—Metaphosphates—Ammonium phosphates—Calcined and other manufactured phosphates—Basic slag, grades, production.

ALTHOUGH the fertilizing effect of bones, in common with most other substances of animal origin, had been known in an empirical way for a very long time, the efficacy was generally put down to the oil they contained, and it was only at the close of the eighteenth century that attention became fixed on the phosphoric acid.

Lord Dundonald, in his *Treatize on the Connection of Agriculture with Chemistry*, published in 1795, had arrived at a very sound perception of the case. When treating of phosphate of lime, he writes that it "is contained in animal matters, such as bone, urine, shells, etc., in some sorts of limestone, and in vegetable substances, particularly in the gluten, or the vegeto-animal part of wheat and other grain. It is a saline compound, very insoluble. There is reason to believe a very considerable proportion of this nearly insoluble salt is contained in most fertile soils. . . . These alkaline phosphates (potash and soda) will be found to promote vegetation in a very great degree; the substance of which they are composed, viz., alkaline salts and phosphoric acid, are found in the ashes of most vegetables." Again, Kirwan writes in 1796 about the constituents of plants: "Phosphorated calx is found in greatest quantity in wheat, where it contributes to the formation of animal gluten. . . . Hence the excellence of bone ashes as a manure for wheat. . . ." Finally, de Saussure, in his *Recherches Chimiques sur la Végétation*, published in 1804, writes: "Le phosphate de chaux contenu dans un animal, ne fait peut-être pas la cinq centième partie de son poids: personne ne doute cependent que ce sel ne soit essentiel à la constitution de ses os. J'ai trouvé ce même sel dans les cendres de tous les végétaux où je l'ai recherché, et nous n'avons aucune raison pour affirmer qu'ils puissent exister sans lui." This opinion was repeated by Davy, and adopted and disseminated by Liebig, by which time various other experimenters had reached the conclusion that the mineral and not the organic matter contained in bones was their chief fertilizing constituent.

Since the phosphatic manures which are of most practical importance are phosphates of calcium, it is necessary to discuss these compounds a little before proceeding further. The phosphatic material which is most widely disseminated, occurring in all the primitive crystalline rocks and occasionally found massive, is the true crystalline mineral apatite, $Ca_{10}(PO_4)_6F_2$, in which the fluorine atom may be wholly or partially replaced by chlorine. This is a definite crystalline compound, the undoubted source of all the other compounds of phosphoric acid, but being very hard and difficult of solution in acids it is little used in manure-making.

The Phosphates of Calcium

The phosphate which is taken as the starting point in manures is tri-calcium phosphate, $Ca_3(PO_4)_2$, but all phosphatic manures are bought and sold in terms of "phosphoric acid," P_2O_5. There are 142 parts of P_2O_5 in 310 parts $Ca_3(PO_4)_2$ or 45·8 per cent.

The term "phosphoric acid" is wrongly used in the fertilizer trade for the compound P_2O_5, the correct chemical name for which is phosphorus pentoxide. This misnomer is, unfortunately, almost universally used and supported by definition in the Fertilisers and Feeding Stuffs Act, 1926.

Phosphorus pentoxide is formed when the element phosphorus burns in a plentiful supply of oxygen, and dissolves in water to form metaphosphoric acid.

$$P_2O_5 + H_2O = 2HPO_3$$

On heating with water, this is converted into orthophosphoric acid

$$HPO_3 + H_2O = H_3PO_4$$

Both acids form salts with bases, but the latter has three replaceable hydrogen atoms and so can form three classes of salts in which a basic radical replaces one, two or all three of the hydrogens. The neutralization with sodium hydroxide may be regarded as taking place in three stages as the following equations show.

(1) $H_3PO_4 + NaOH = NaH_2PO_4 + H_2O$
mono-sodium phosphate

(2) $H_3PO_4 + 2NaOH = Na_2HPO_4 + 2H_2O$
di-sodium phosphate

(3) $H_3PO_4 + 3NaOH = Na_3PO_4 + 6H_2O$
tri-sodium phosphate

The corresponding calcium salts are mono-calcium phosphate $Ca(H_2PO_4)_2$ which is soluble in water, di-calcium $CaHPO_4$ and tri-calcium phosphate $Ca_3(PO_4)_2$ which are insoluble in water. The last mentioned is commonly referred to simply as calcium phosphate.

When tri-calcium phosphate is treated with an equivalent amount of sulphuric acid, phosphoric acid and calcium sulphate are produced.

$$Ca_3(PO_4)_2 + 3H_2SO_4 = 2H_3PO_4 + 3CaSO_4$$

But if only enough acid to displace 2 of the calcium radicals is used, a mixture of mono-calcium phosphate and calcium sulphate, known as superphosphate, is produced

$$Ca_3(PO_4)_2 + 2H_2SO_4 = Ca(H_2PO_4)_2 + 2CaSO_4$$

Mono-calcium phosphate, soluble in water, readily "reverts" to an insoluble form when treated with a calcium compound,

e.g. $Ca(H_2PO_4)_2 + Ca(OH)_2 = 2CaHPO_4 + 2H_2O$

Di calcium phosphate has sometimes come on to the market as a fertilizer; it is a perfectly stable compound and, as it is produced by chemical precipitation, is in a fine state of division and readily becomes available to the plant.

Bone Manures

Of the phosphatic manures, the earliest and for a long time the only ones to be employed on a large scale were those derived from bones. It would be impossible to attribute the discovery of the fertilizing value of bones to any individual; in common with all other waste materials of animal origin, they were probably tried and appreciated by numbers of people in all ages and countries; they are mentioned by Blithe in 1653, Evelyn in 1674, and Worlidge in 1668, and by the close of the eighteenth century their use was becoming common in the neighbourhood of all the great towns. Arthur Young mentions the use of the waste from the making of knife-handles near Sheffield, and again enumerates bones as one of the substances the Hertfordshire farmers were in the habit of bringing back from London when their carts had been delivering hay or grain.

It was in the early years of the nineteenth century, however, that the demand began to grow and it received a considerable impetus from the introduction, probably first of all in Yorkshire, of machines for reducing the bone into half- or quarter-inch fragments, or even into powder. By 1815 the home supply was proving insufficient, and bones began to be imported from the Continent in rapidly increasing quantities until nearly 30,000 tons per annum were brought in, chiefly from Europe—a demand which is said to have resulted in the ransacking of many of the battlefields. In this connection a characteristic outburst of Liebig's has often been quoted: "England is robbing all other countries of their fertility. Already in her eagerness for bones, she has turned up the battlefields of Leipsic, and Waterloo, and of the Crimea: already from the catacombs of Sicily she has carried away the skeletons of many successive generations. Annually she removes from the shores of other countries to her own the manurial equivalent of three million and a half of men, whom she takes from us the means of supporting, and squanders down her sewers to the sea. Like a vampire she hangs upon the neck of Europe, nay, of the whole world, and sucks the heart blood from nations without a thought of justice towards them, without a shadow of lasting advantage to herself!"

For a time the importations fell off, but with the growth of the artificial manure trade and the opening up of India and South America as sources, the amounts introduced increased enormously. Since the discovery of basic slag and the cheapening of mineral phosphates, they have steadily decreased. In 1906 the imports amounted to 42,600 tons, in 1936 to 11,200 tons, in 1948 to 8,400 tons.

Fresh bone has the following percentage composition: water 50, cartilage protein (collagen) 20, fat 4, mineral matter 26. The mineral matter consists of about 85 per cent. calcium phosphate, 14 per cent. calcium carbonate, and 1 per cent. magnesium phosphate, $Mg_3(PO_4)_2$; it may be regarded as a carbonate apatite. If the bone is allowed to soak for a time in dilute acid, the mineral framework is dissolved; if the bone is strongly heated, the protein and fat are burned off leaving bone ash. Commercial bones generally contain a smaller proportion of water than fresh bones. In preparing bones, they are first treated with steam under a pressure of 15 to 20 lb. in order to melt and remove the fat, which is sold as tallow or used for soap-making. In some cases the fat is extracted even more thoroughly by the action of a fat solvent. The extracted or steamed bones thus obtained contain from 3 to 5 per cent. of nitrogen and 43 to 50 per cent. of calcium phosphate, and are ground down into "bone meal." A really fine powder is, however, rarely obtained, because the cartilage interferes materially with the disintegration unless special methods are employed. It is this crushed material which is sometimes treated with sulphuric acid in the manufacture of "dissolved" or "vitriolized" bones. In factories making glue, the cleaner bones are picked out and, after the fat extraction, are broken up and steamed afresh at a much higher pressure and temperature, 50 to 60 lb. to the square inch, by which process the collagen becomes converted into gelatine. The solution is concentrated and allowed to set, when it becomes glue; the bone residue, which now contains only about 1 per cent. of nitrogen but 55 to 60 per cent. of calcium phosphate, is ground and sold as "steamed bone flour." Owing to the removal of the cartilage, this material can be finely ground to a dry friable powder. At one time steamed bone meal was often treated with acid to obtain a "soluble bone compound" which was sometimes mixed with nitrogenous material.

Bone meal is a somewhat gritty powder with a strong and distinctive smell, and it should contain not less than 45 per cent. of calcium phosphate. The percentage of nitrogen is more variable: good fresh English samples sometimes show 5, but 4 per cent. is good, and Indian samples which have been much weathered and are a little decayed may fall to 3 or even lower; this nitrogen is not present in a very active form, the cartilage being slow to decompose in the soil. Of the phosphates in bones about one-half can be dissolved on shaking up 1 gram of the bone meal with 1 litre of 1 per cent. citric acid solution, which indicates that the phosphate is easily available. However, bone meal is rather slow acting, probably because of its comparative coarseness and the consequent small surface of the particles offered to the solvent action of the soil water. It was, however, bones in their even coarser form—merely roughly broken, sometimes by hand on the farm—which built up the fertility of much English land, as, for instance, the famous dairy pastures of Cheshire, which were made during the early years of the nineteenth century. Large dressings of bones were employed—a ton or more per acre—and the application was expected to last for twenty years, little return being obtained during the first year or two; for this reason the landlord contributed freely to the cost of boning, even if he did not pay for it entirely. The pastures improved steadily after the dressing of bones; in particular, such a growth of white clover was encouraged that some farmers actually suspected the manure to contain clover seed, a belief which reappeared fifty years or more later when basic slag first began to be used on clay pastures.

Steamed bone flour, because of its freedom from cartilage and finer grinding is a much more readily available form of phosphate than bone meal. It was shown to be particularly valuable for the turnip crop on light sandy soils, and is an excellent conditioner for mixed fertilizer compounds.

Dissolved bones or bone superphosphate was a popular manure at one time. It contained 8 or 9 per cent. P_2O_5 of which about half was readily soluble, and about 3 per cent. of nitrogen. But it was a rather sticky brown product difficult to sow. The other bone compounds were often complex mixtures of mineral phosphates and superphosphate, made from steamed bone flour, and certain animal waste products containing nitrogen. They are seldom offered for sale now.

In 1938, products from bone accounted for only about 3 per cent. of the total P_2O_5 used in fertilizers. It is now much less.

Guanos

In certain dry areas of the world are to be found islands which are breeding-places of sea-birds and on which deposits of excreta and animal remains may accumulate to a considerable depth. For example, on some islands off Peru the bed of guano was about 100 feet thick and large quantities were exported as manure in the nineteenth century. Similar deposits occurred in various parts of the West Indies and off the coast of South Africa where the collection is now controlled. The composition of the guano varied greatly according to the rainfall. Samples might contain from about 14 per cent. nitrogen and 20 per cent. calcium phosphate to about 2 per cent. nitrogen and 60 per cent. calcium phosphate. Further details are given in Chapter IX. In 1938, guano accounted for only about 0·7 per cent. of the total P_2O_5 used in fertilizers. With weathering, the nitrogen is lost and the phosphate becomes incorporated with the underlying rock. On Curaçao in the West Indies, and on Ocean and Nauru Islands of the Pacific, the coral rag has become permeated with guano leachings. In the Pacific islands the nitrogen has been entirely removed to leave a deposit consisting of 85 to 88 per cent. calcium phosphate. Such deposits can no longer be regarded as guano and are generally considered as mineral phosphates.

Mineral Phosphates

With the recognition that the fertilizing value of bones lay in the calcium phosphate they contained, which we may conclude had become the accepted opinion about 1840, attention began to be directed to mineral sources of phosphate —apatite and phosphorite, the existence of which had long been known to mineralogists. Acting on an analysis of Proust's, Professor Daubeny and Captain Widdrington made an expedition in 1843 to Estremadura, in Spain, to find a bed of phosphatic rock there reported. They discovered the deposit and secured enough of it for a few field experiments in England in the following year, but difficulties of transport prevented anything more than small quantities being exported until a much later period. The immediate demand for such material was satisfied by the discovery in 1845 by Professor Henslow of the bed of coprolites lying at the base of the Greensand near Cambridge.

Origin.—As previously mentioned, all phosphate deposits are to be traced from eruptive apatite rock of general composition, $3Ca_3(PO_4)_2.CaX_2$, where X is OH (hydroxyl) or F (fluorine). Some phosphorus may also be found in primitive rock as ferrous phosphate $Fe_3(PO_4)_2$. Secondary deposits are found in the fissures of eruptive rocks, e.g. apatite, free from chlorine or fluorine by lixiviation, is found in crevices in basalt or dolerite, such as the phosphorite deposits in Spain and France. The disintegration and solution of apatite followed by precipitation leads to deposits of meadow ore or vivianite, $Fe_3(PO_4)_2$, but the phosphorus of iron ores is mostly of organic origin.

Plants have provided the phosphorus to form animal skeletons. The accumulations of carcasses or animal waste in limited spaces such as the stranding of whales or the drifting of animalcules into shallow bays or deposits of shell fish, followed by the formation of phosphorite with clay mud, account for deposits of animal origin. Coprolites may be phosphorite nodules formed by the petrifaction of animal excrement, or they may have arisen through the deposition of phosphates, dissolved from bones, around stones and organic remains. Sometimes they are marine deposits laid down in shallow waters, the grading according to size and density being carried out on the rubble slope. The land pebble phosphates represent old river terraces. The Florida land pebbles, for example, are mainly small granules consisting of about 75 per cent. calcium phosphate and 7 per cent. calcium carbonate. The soft phosphate rock to be found interstratified with sandstone or shale, as in North Africa, was probably formed through disintegration of fossil remains. The North Africa deposits are actually derived from marine animal skeletons of Eocene Age and decaying vegetable matter. The Tennessee and Kentucky deposits have been formed from the decomposition of phosphatic dolomite or limestone.

The older the deposit the harder the material becomes. Microscopic and X-ray examinations have shown that phosphate rocks consist in the main of fluorine-apatite $Ca_{10}(PO_4)_6F_2$. Rocks of older geological periods also contain hydroxy-apatite $Ca_{10}(PO_4)_6(OH)_2$ which is now considered to be of greater importance than pure tricalcium phosphate. N. African rock consists largely of a carbonate-fluorapatite $Ca_{10}(PO_4.CO_3)_6F_{2-3}$ whilst the Ocean and Nauru deposit is chiefly a carbonate-hydroxy-fluorapatite, of approximate formula $Ca_{10}(PO_4,CO_3)_6(OH,F)_{2-3}$

Production.—The English coprolites—small rounded nodules of impure calcium phosphate, mixed with fragments of bone and shell, shark's teeth, etc.—occur at various horizons in the newer secondary and tertiary rocks, e.g. at the base of the Upper Greensand and at the base of the Gault, and in the Crag, where it rests upon the London Clay. Shortly after their discovery these deposits began to be worked for coprolites at various places in Suffolk, near Cambridge, and at Potton in Bedfordshire. The output reached 50,000 tons or so in the early eighties of the last century, but then rapidly declined owing to the opening up of so many cheaper foreign deposits, and has of late years entirely ceased. The coprolites formed hard dark-coloured nodules, which were ground down to a grey powder containing from 40 to 60 per cent. of calcium phosphate, about 10 per cent. of calcium carbonate, and 3 per cent. of calcium fluoride. Though occasionally applied directly to the land in a ground form, they were almost wholly used in the manufacture of superphosphate.

Similar to the English coprolites were the phosphates obtained from France, Germany, and Belgium, where they occur in the secondary and tertiary formations on a more important scale than do the similar deposits in England. Of these materials the most important were the Lahn phosphates, extensively worked for some time after their discovery in 1864, the Belgian phosphates worked near Mons, with 45 to 60 per cent. of calcium phosphate, and the Somme phosphates, of which extensive deposits were found in the north of France, and formed an important source of supply to the manure market about 1890.

The Lahn phosphates fell out of favour because of the large amount of iron and aluminium they contained, the Belgian phosphates became of too low grade, but the Somme phosphates remained valuable because they contained in the better grades 70 per cent. or so of calcium phosphate, and only 1 to 2 per cent. of oxides of iron and aluminium. They also yielded a very dry and friable superphosphate, and were useful for mixing with the Florida phosphates before treatment with acid. These coprolitic phosphates, however, attain their greatest development in Florida. There in many places the subsoil is a sandy deposit full of coprolitic pebbles, which can readily be separated by screens or washing; the beds of the rivers and creeks, again, are wholly composed of the same pebbles, which are recovered by dredging. The fine material is recovered by flotation methods in which the fine particles of phosphates are floated off from the sand by being suspended in a film of soap and oil. On concentration, the product contains about 70 per cent. calcium phosphate. The river deposits were particularly valued in Great Britain for superphosphate making, because they contained from 60 to 77 per cent. of calcium phosphate and were specially free from iron and aluminium.

All these phosphate deposits are now yielding to the competition of the great deposits of phosphate rock discovered in Northern Africa and exported in immense quantities from Algeria, Morocco and Tunis. The phosphate bed appears to stretch right across the continent and Egyptian rocks are also being worked in considerable amounts. The most important of the phosphate mines in North Africa occur in the province of Constantine in the district of Tebessa, from whence they extend into Tunis, near Gafsa. The rock is generally at the base of the Eocene system, and occurs in strata that may be 8 or 10 feet thick and contain as much as 60 per cent. of calcium phosphate, which may be raised to 70 per cent. by careful selection.

The beds are extensive and mining is done by tunnelling, is mainly horizontal and is rarely deep. The phosphate is soft in character and before export is dried first by exposure to the air in summer and then in electric furnaces. These African phosphates contain less than 2 per cent. iron and aluminium, and are the chief material for the manufacture of superphosphate in this country.

Nauru accounts for a large proportion of the Pacific Islands' phosphate production and the deposit here consists in the main of a friable layer of variable thickness which is excavated, crushed and dried. On account of the coral reef round the island, loading presented special difficulties which have been overcome by the construction of a 200-foot-long cantilever, supporting belt conveyors which can transfer the phosphate direct to ships lying in deep water. The material is of very high quality, containing from 83 to 88 per cent. calcium phosphate.

The fourth great source of phosphate lies within the Arctic Circle on the Kola peninsula of U.S.S.R. The discovery of this high-grade mineral apatite, containing about 66 per cent. calcium phosphate, was made about 1930. The material is a very hard primary rock formation in contrast to the soft secondary rock phosphates and requires to be blasted, but very extensive mining operations have already been carried out.

Of all the phosphate produced, it is estimated that about 7 per cent. is required in the metallurgical and numerous chemical industries, whilst the remainder is used in agriculture. Nearly 80 per cent. was converted into superphosphate in 1938 and just about 10 per cent. was used directly as a fertilizer. With certain fluctuations, the output has risen from one to 23 million tons in the 64 years between 1888 and 1952. The following table gives a summary of the production of phosphate rock in 1949, with the corresponding figures for P_2O_5, all in millions of metric tons.

The average percentage of P_2O_5 in the rock was 32·2.

TABLE 13.—WORLD PRODUCTION OF ROCK PHOSPHATE.

	Phosphate Rock		P_2O_5	
U.S.S.R. . . .	2·20		0·69	
Total for Europe . .	——	2·42	——	0·71
Morocco . . .	3·15		1·04	
Tunis . . .	1·86		0·52	
Total for Africa . .	——	6·10	——	1·87
U.S.A. . . .	9·25		3·03	
Total for America . .	——	9·68	——	3·13
Ocean and Nauru . .	0·78		——	
Total for Pacific . .	——	0·85	——	0·33
World Total . . .		19·38		6·14

In the same year, the British Isles imported and consumed approximately 1 million tons.

Latest figures for production in U.S.A. show a remarkable increase to 13 million tons of rock in 1953, with a consumption of 2·54 million tons of P_2O_5; the output from Morocco has also risen quickly to 4 million tons in 1953.

Superphosphate

In the early years of the nineteenth century, long prior to the introduction of superphosphate as a manure, the existence of a soluble calcium phosphate produced by the action of sulphuric acid upon bone ash was a matter of common chemical knowledge, and the composition of this and the other phosphates had been studied accurately by Berzelius. The application of this knowledge to agriculture and the introduction of superphosphate as a manure began about 1840. The first published suggestions of the kind were by Escher of Brunn, in 1835, who considered that the fertilizer action of bone meal would be speeded up by treatment with acid; by Ridgeway in a pamphlet on bone manure dated 1839, in which the first recorded use of the term superphosphate occurs; by Liebig, in his Report to the British Association, published in September, 1840, *The Chemistry of Agriculture and Physiology.* Writing of bones as a manure and the necessity of their being finely divided, he goes on: "The most easy and practical mode of effecting their division is to pour over the bones, in a state of fine powder, half of their weight of sulphuric acid, diluted with three or four parts of water; and after they have been digested for some time to add 100 parts of water, and sprinkle this mixture over the field before the plough. In a few seconds the free acids unite with the bases contained in the earth, and a neutral salt is formed in a very fine state of division. Experiments instituted on a soil formed from a *grauwacke*, for the purpose of ascertaining the action of manures thus prepared, have distinctly shown that neither corn nor kitchen garden plants suffer injurious effects in consequence, but that, on the contrary, they thrive with much more vigour." Liebig then adds: "In the manufactories of glue, many hundred tons of a solution of phosphates in muriatic acid are yearly thrown away as being useless. . . . It would be important to examine whether this solution might not be substituted for the bones."

In May, 1842, J. B. Lawes was granted a patent for making superphosphate, from the specification of which the following extract has been made: "Whereas bones, bone ash, and bone dust and other phosphoritic substances have been heretofore employed as manures, but always, to the best of my knowledge, in a chemically undecomposed state, whereby their action on the soils to which they have been applied has been tardy and imperfect. And whereas it is in particular well known that in the case of a large proportion of the soils of this country, the application of bone dust is of no utility in producing crops of turnips on account of the slow decomposition of the bone dust in the soil, and the consequent exposure of the young plant for a long period to the ravages of the turnip fly. Now, the first of my said improvements consists in decomposing, in manner following, the said bones, bone ash, bone dust, and other phosphoritic substances. Previous to using them for the purposes of manure, I mix with the bones, bone ash or bone dust, or with apatite or phosphorite, or any other substance containing phosphoric acid, a quantity of sulphuric acid just sufficient to set free as much phosphoric acid as will hold in solution the undecomposed phosphate of lime."

Subsequently, on becoming acquainted with Liebig's published suggestion, Lawes amended his patent by disclaiming all references to bone and bone products, and confining it to "apatite and phosphorite, and other substances containing phosphoric acid." Following on his patent, Lawes began the manufacture of superphosphate on a commercial scale, establishing a factory at Deptford and using for the purpose at first bone ash and later the crag coprolites from Suffolk, to which Henslow's paper in 1845 had attracted attention. The first advertisement appears in the *Gardener's Chronicle and Agricultural Gazette* in 1843, the price being 4s. 6d. per bushel. From the dates of Liebig's book and Lawes' patent, Liebig has generally been regarded as the inventor of superphosphate, and even though Lawes was able to take out a patent for making it from mineral phosphates instead of from bones, the idea is generally set down to Liebig. Owing, however, to actions for infringement of his patent brought by Lawes in 1853, the steps leading up to Lawes' patent are on record, and it is seen that he arrived at the idea of making superphosphate and had tried it experimentally on a considerable scale, prior to Liebig's publication.

In his proof of evidence, Lawes, after describing his fitting up of a laboratory and taking over of the Rothamsted estate in 1834, stated that during 1836, 1837, and 1838, he used considerable quantities of bone dust on his farm for the purpose of manuring his turnip crops, and finding that it produced no good effects, and knowing that upon other soils its properties as a manure were very great, he commenced a series of experiments with bones and mineral phosphate of lime decomposed by sulphuric and other acids, applied to the most important agricultural plants of the farm. These experiments were in 1839 conducted on a small scale by plants in pots and by manuring a certain number of plants in a field. The result of these small experiments was most remarkable with turnips and, in 1840, he used the superphosphate on half an acre or more of that crop. In 1841 the results were so far advanced that he used about 20 tons of superphosphate. He was prepared to have taken out a patent in 1841, but was persuaded by his friends not to do so, as they objected to his embarking in any mercantile or business occupation (Lawes was then only twenty-six years of age). Lawes further stated that he had not noticed Liebig's recommendation until after the date of his patent, whereupon he disclaimed the use of bone substances included therein. He also declared that he began his manufacture with bone ash, apatite and phosphorite being unobtainable commercially, though in 1843 he imported several tons of the Spanish phosphorite from Estremadura, and early in 1845 began his inquiries if coprolites could be obtained cheaply from the Eastern counties. Gilbert also gave evidence that the manufacture and use of superphosphate on a large scale prior to the date of the patent were matters of common knowledge at Rothamsted when he came there in 1843. Putting aside the fact that Lawes was the first man to make superphosphate a practical possibility, there is thus no doubt that he arrived at the idea of the importance of a soluble phosphatic manure quite independently of Liebig, and had tested the idea on a working scale before taking out his patent. The only other point of interest in this early history of superphosphates is that Sir James Murray, of Dublin, was granted a patent on the same day as Lawes, in which he suggested as manure, phosphoric acid prepared by treating phosphorite with sulphuric acid. Murray's object, however, was to generate carbonic acid in the ground, and his patent is for a means of "mechanically fixing and solidifying mineral acids"—sulphuric, muriatic, nitric, and phosphoric—by mixing them with absorbent matter like bran, sawdust, peat, etc., the phosphoric acid thus coming in incidentally only, and not for its own nutritive value to plants. A few years later, to avoid any question of priority that might arise, Lawes purchased Murray's patent, and amended it by disclaiming everything but the manufacture of a manure by the treatment of phosphorite with acid.

The early superphosphate thus manufactured by Lawes and sold at about £7 a ton, was a mixture of soluble and insoluble phosphate derived from coprolites or from guanos, mixed with animal substances and ammoniacal salts. Way, in 1851, gives analyses ranging from 3·24 to 0·12 per cent. of nitrogen, soluble calcium phosphate from 18·5 down to 1·6 per cent., insoluble from 28·3 down to 0·06 per cent.

Superphosphate was manufactured only in Great Britain until 1854, by which time the production had risen to about 30,000 tons per annum. In 1905 the world production had reached 6 million tons of which over two-thirds was in Europe. The development has suffered seriously at various times but, about 1925, it became general all over the world, and now accounts for more than 70 per cent. of the total phosphorus in fertilizers of which some world statistics are given below.

TABLE 14.—CHIEF PRODUCERS AND CONSUMERS OF PHOSPHATES, 1952–53. (In millions of metric tons, P_2O_5.)

	Production.	Consumption.
France	0·42	0·45
Germany	0·43	0·46
United Kingdom	0·33	0·36
Netherlands	0·12	0·12
Italy	0·36	0·34
Belgium	0·21	0·08
U.S.A.	2·19	2·16
Japan	0·25	0·25
Australia	0·35	0·35
New Zealand	0·13	0·14
World Total	5·98	6·09

(Exclusive of U.S.S.R. and of figures for Rock Phosphate.)

Manufacture.—In the early days superphosphate was made by hand, rather like mortar, by mixing sulphuric acid with calcined bones or ground coprolites in a pit. Then the mixing den was introduced, capable of holding about 50 tons of the mixture, but still emptied by hand. Various types of mechanical mixers were introduced to improve the process and about 1910 there followed mechanical dens. In these, provision is made for the den to move forward against revolving knives, or for a cutter to move into the den so that the mass of superphosphate can be reduced to powder form and removed on conveyor bands. Gradually more and more mechanism has been introduced into the plants leading eventually to dens which produce continuously instead of intermittently. The acid and ground phosphate are continuously mixed in measured quantities and the semi-solid mass allowed about 3 hours to complete the chemical action and solidify before reaching the excavator section. In one case the den consists of a kind of slowly-moving box, the movable parts consisting of wooden slats on an endless chain.

In another, the den is in a large revolving cylinder which takes about 3 hours to make one revolution so that the freshly-mixed mass deposited in the cylinder takes that time to move round to the excavation knives. Yet another den consists of a large circular tube revolving like a wheel and so timed that the finished product is continuously removed. An entirely different type of process makes use of a rotating autoclave in place of a den, into which the acid and phosphate are forced under pressure. The chemical reaction is speeded up by steam heating and reduced pressure, so that the autoclave may be cleared after an hour when the product is hard and dry and granular in character. In short, all these modern processes of manufacture have been designed to increase production capacity and improve the quality of the superphosphate. The importance of a low moisture content will be readily appreciated, and a good deal of attention has been given in recent years to methods of drying. The modern tendency has been towards the production of fertilizers—both straight and compound—in granular form which commonly involves the handling of a paste from which the water has eventually to be driven off. The rotary drum type of drier has proved to be a valuable part of the equipment in a fertilizer factory.

Composition.—The object of manufacturing superphosphate is to convert as much as possible of the insoluble tri-calcium phosphate in mineral phosphate into a more soluble form. Throughout the world the quality of superphosphate is based on its P_2O_5 content, but of the eight largest consumers, United Kingdom, Germany, Australia and Japan require a guarantee of the percentage of P_2O_5 soluble in water, whilst France, Italy and U.S.A. specify the P_2O_5 soluble in water or in 2 per cent. ammonium citrate. The availability of phosphate is a controversial subject and is discussed later; but, as far as superphosphate is concerned, it is essentially a question of converting the tri-calcium phosphate into the mono-calcium phosphate by replacing two-thirds of the calcium by hydrogen. The reaction in its simplest form may be written:

$$Ca_3(PO_4)_2+2H_2SO_4+4H_2O=Ca(H_2PO_4)_2+2CaSO_4.2H_2O$$

the superphosphate consisting of the mixture of approximately 2 parts by weight of mono-calcium phosphate and 3 parts of gypsum. Even the highest grades of phosphate rock, however, contain certain amounts of such compounds as calcium carbonate and fluoride, ferric phosphate and aluminium silicate, which are also decomposed by the sulphuric acid to form sulphates and liberate water, carbon dioxide and hydrofluoric acid as gases. The weight of superphosphate produced is therefore less than the weight of the original mixture of phosphate and acid. About 11·7 cwt. of phosphate and 10·5 cwt. of 70 per cent. acid are required to give 1 ton of superphosphate. The manufacturer must calculate the amount of acid required, knowing the composition of the rock phosphate to be treated. The complete composition of superphosphate will vary according to the constitution of the rock phosphate used and method of manufacture, but the percentage composition of a typical freshly-made superphosphate might be as follows:

10 to 14 of water (free and combined)
25 of mono-calcium phosphate equivalent to 15 per cent P_2O_5
5 of orthophosphoric acid equivalent to 3·6 per cent. P_2O_5
50 of gypsum and

a variety of other salts and possibly some organic matter. This mixture would contain 18·6 per cent. water-soluble P_2O_5 together with, perhaps, 0·5 per cent. P_2O_5 in di-calcium phosphate, insoluble in water but soluble in citrate solution, and 0·5 per cent. P_2O_5 in undissolved tri-calcium phosphate insoluble in either water or ammonium citrate.

The finely-ground mineral phosphate is mixed with the calculated amount of 70 per cent. sulphuric acid. There is a considerable rise in the temperature of the semi-liquid mixture with evolution of gases, which must be trapped, and then solidification takes place. About 2 to 4 hours are allowed for the reaction to complete itself in the den, then the product is cut out and usually stored in heaps for some time to cure before being ground and bagged. Various grades are manufactured depending upon the percentage of P_2O_5 in the rock available. The percentage of water-soluble P_2O_5 in the superphosphate will be about half of the percentage of total P_2O_5 in the raw material if the latter does not contain more than about 1·5 per cent. of iron and aluminium oxides. A high content of iron and aluminium is most undesirable for they not only consume sulphuric acid but form iron and aluminium phosphates which are quite insoluble and so reduce the water-soluble P_2O_5. On the other hand, the presence of a small amount of carbonate (up to 5 per cent.) in the rock is helpful, for the carbon dioxide liberated makes the superphosphate porous and facilitates drying. Fluorine is an invariable constituent of phosphate rock; the presence of fluor-apatite reduces the availability of phosphorus to the plant and this compound is broken down by the acid. At one time the toxic hydrofluoric acid evolved was a most undesirable feature in superphosphate manufacture, but the value of fluosilicates as insecticides now makes it worth while to absorb the gas.

The bulk of ordinary superphosphate sold contains 18 to 20 per cent. P_2O_5 soluble in water. In this country the 18 per cent. grade is probably in greatest demand, although higher grades are sold in certain districts. There is no question of the popularity of superphosphate as the following table of forms of phosphatic fertilizers shows.

TABLE 15.—WORLD CONSUMPTION OF P_2O_5 FOR AGRICULTURAL PURPOSES, 1953.

Source.	Millions of Tons P_2O_5.	Per Cent. of Total.
Superphosphate	4·38	73
Basic Slag	0·78	13
Ammonium Phosphate	0·24	4
Other Fertilizers	0·60	10
Total	6·0	...

Reactions.—The reason for the value of superphosphate does not lie entirely in its content of phosphate soluble in water, for in presence of calcium compounds the soluble mono-calcium salt is rapidly converted to the di-calcium salt.

$$Ca(H_2PO_4)_2 + CaX = Ca_2(HPO_4)_2 + H_2X$$

This "reversion" probably takes place soon after superphosphate comes into contact with the soil, but the di-calcium compound is precipitated in a state of very fine division and, being soluble in dilute acids, is readily utilized by the plant. Incidentally, certain other phosphatic fertilizers which are insoluble in water but readily dissolved in dilute acids are often as effective as superphosphate in increasing crop yield. But it is important that the di-calcium form should persist because if the reversion goes further,

$$Ca_2(HPO_4)_2 + CaX = Ca_3(PO_4)_2 + H_2X$$

the tri-calcium compound produced may form one of the very insoluble apatites. If the soil is very acid, iron and aluminium salts may react to give the insoluble iron and aluminium phosphates. Hence the importance of regular liming to conserve available phosphate. The phosphate which is stable in soils is probably the hydroxy-apatite $3Ca_3(PO_4)_2.Ca(OH)_2$ whose availability depends upon its physical condition.

Mono-calcium phosphate is an acid salt and consequently can displace exchangeable calcium in the soil by hydrogen and make the soil acid. But the displaced calcium, as shown above, forms an insoluble phosphate and is not lost in the drainage water. More than half of the superphosphate mixture consists of calcium sulphate which is soluble in water, so that another exchange reaction may take place in which calcium displaces the exchangeable hydrogen of the soil and so reduces soil acidity. Obviously the total chemical reaction is complicated. Although an abnormally large dressing of superphosphate does increase the acidity of the soil, dressings of the order of those used in practice have very little effect. The effects of annual applications of about 3 cwt. superphosphate per acre on certain plots for 68 years at Rothamsted, and 47 years at Woburn were actually to increase the percentage of exchangeable calcium in the surface soils.

Concentrated Superphosphate

If rock phosphate is treated with phosphoric acid instead of sulphuric acid there will be no formation of calcium sulphate, and the resulting mixture will be much richer in P_2O_5. There are two common methods of manufacturing phosphoric acid. In the "wet" process a dilute solution of the acid is separated from a phosphate—sulphuric acid mixture, by filtering off the sulphate sludge, and concentrated by evaporation. It can also be obtained by burning the element phosphorus in air and dissolving the resultant pentoxide, P_2O_5, in water; the phosphorus is manufactured by heating phosphate rock with carbon (see below). Because its production takes place in two stages (*a*) preparation of phosphoric acid, (*b*) treatment of rock phosphate with phosphoric acid, the concentrated phosphatic compound is sometimes called "double" superphosphate, but in U.S.A. it is known as "triple" superphosphate because it contains nearly three times as much P_2O_5 as ordinary superphosphate.

The manufacture of "triple" superphosphate was started in Europe at the beginning of the century but little was made until quite recently. Finely-ground phosphate and 78 per cent. phosphoric acid solution are intimately mixed and stored for about 3 months to allow the reaction to be completed. The tough grey mass of material is then broken up and crushed to a suitable fineness for easy distribution. The product of the Tennessee Valley Authority contains 47 to 48 per cent. P_2O_5 soluble in water or in ammonium citrate; 40 per cent. is soluble in water. In the production of phosphoric acid by the calcining process the phosphate is first heated to a temperature high enough to melt the fine particles and form nodules, and then mixed with coke and quartz gravel which acts as a flux. The mixture is heated in an electric furnace to 1,500°C. The carbon reduces the phosphate to give carbon monoxide and phosphorus vapour, and the calcium forms a slag with the silica. The carbon monoxide is used for heating purposes, the phosphorus is condensed to a liquid, the hot slag is broken down by a stream of water under pressure to a fine sand which contains nearly 50 per cent. CaO and has proved to be a valuable liming material. The phosphorus is burned to produce P_2O_5 which is dissolved in a spray of water to give a syrup containing 85 per cent orthophosphoric acid, H_3PO_4. The advantages claimed for this method in Tennessee are that phosphate of grades unsuitable for sulphuric acid treatment can be used and, since electric power is available at the mines, only the phosphorus need be transported to the fertilizer plants. But, where sulphuric acid is cheap and the difference in cost of transport between 2½ tons of ordinary superphosphate and 1 ton of triple superphosphate is less than the extra cost of the double process, ordinary superphosphate will probably continue to be manufactured as at present.

Metaphosphates

A further recent development in the manufacture of phosphatic fertilizers has been the treatment of lump rock phosphate with vapours of P_2O_5. The tri-calcium phosphate is converted into calcium metaphosphate, a salt of metaphosphoric acid,

$Ca_3(PO_4)_2 + 2P_2O_5 = 3Ca(PO_3)_2$, calcium metaphosphate, which is obtained in the form of a glass and is crushed and screened. The pure compound contains the equivalent of 71 per cent. P_2O_5, but the commercial product contains various impurities and the total phosphorus content expressed as P_2O_5 amounts to about 64 per cent., the bulk of which is soluble in ammonium citrate. It is probable that metaphosphates are hydrolysed in water to give the PO_4 ion. If a potassium salt is introduced during the reaction, potassium metaphosphate is produced. The anhydrous salt, KPO_3, contains the equivalent of 60·1 per cent. P_2O_5 and 39·8 per cent. K_2O, and is soluble in water, but the commercial product is hydrated and its solubility depends upon the presence of other salts. At present, work with these products is still in the experimental stage, but there would seem to be considerable scope for vitreous fertilizers containing definite proportions of plant nutrients which might dissolve and become available throughout the growing season.

Ammonium Phosphates

Many years ago it was suggested that ammonia might be added to superphosphate to improve the latter's mechanical condition and storage qualities. About 3 per cent. ammonia may be added without seriously decreasing the amount of soluble P_2O_5 and the product is called ammoniated superphosphate. The production of mono-ammonium phosphate, $NH_4.H_2PO_4$, has, however, been of much greater importance. Pure mono-ammonium phosphate may, of course, be prepared by adding the required amount of ammonia to phosphoric acid and contains 12·2 per cent. nitrogen and 61·7 per cent. P_2O_5. Another method of manufacture has already been described (Chapter III) in which rock phosphate is treated with ammonium sulphate and sulphuric acid. There is also a considerable production of di-ammonium phosphate $(NH_4)_2HPO_4$.

Mono-ammonium phosphate is a stable, crystalline and non-hygroscopic salt completely soluble in water, and is undoubtedly the most important of all the concentrated fertilizer salts. It is estimated that about 225,000 tons were produced in Europe in 1938, mainly as a basis for concentrated compound fertilizers. The commercial product contains about 12 per cent. N and 50 to 54 per cent. P_2O_5. By combining it with an equal weight of ammonium sulphate the figures for N and P_2O_5 become 16 and 27; by combining one part of the compound with 2 parts ammonium sulphate the figures become 18 and 18. Obviously the ratio of N to P_2O_5 may be altered at will to obtain mixtures suitable for different crops and soil conditions, and such mixtures will always contain from 2 to 3 times as much plant food (calcium excepted) as the same total weight of ammonium sulphate and ordinary superphosphate. But potassium salts may also be incorporated to give various ratios of N: P: K, and sometimes a little mineral phosphate is added so that calcium and insoluble P_2O_5 are also provided. Several such "complete" fertilizers are produced in this country and the percentage composition of three of them is given to show the variety available.

N	9	12	7·5
Water-Soluble P_2O_5	9	12	26·0
Insoluble P_2O_5	...	...	6·0
K_2O	15	15	7·5

They are all produced in granular form and there is reason to suppose that a considerable expansion in their use will take place since they economize in cost of transport and in labour of handling per unit of plant food.

Calcined and Other Manufactured Phosphates

In the search for methods of converting the phosphate in phosphate rock into a more available form without using sulphuric acid, a good deal of attention has been given to calcination. The idea is an old one and received some impetus with the discovery of the value of basic slag as a phosphatic fertilizer. It is really a question of breaking down the fluorapatite by fusing it in presence of some suitable agent. It has generally been found that suitable agents are either cheap, like chalk and silica, but have to be used in such large amounts that the fusion is costly, or are rather expensive like sodium carbonate. The greater the proportion of alkali used the lower are the temperatures required for the conversion, and steam has been found to help the reaction. A temperature of about 1,400°C is necessary, and the elimination of the fluorine presents some difficulties. Fusion of the rock phosphate with olivine, a magnesium silicate, has recently been reported to have met with some success. Several products have been marketed, of which German Rhenania phosphate with 25 to 28 per cent. P_2O_5 is probably the best known. The P_2O_5 in these substances is not soluble in water but soluble in citrate solutions and available to plants; it may be, therefore, that such products will compete with superphosphate only in those countries where "water-soluble" and "citrate-soluble" P_2O_5 are regarded as being of equal value.

This also applies to certain other products which are richer in P_2O_5 than superphosphate, but in which only a portion of the available P_2O_5 is soluble in water. In one type the superphosphate is made with only half the acid normally used (Kotka phosphate in Finland), in another the ordinary superphosphate is mixed with an equal weight of ground rock phosphate ("Semsol" in Ireland). Yet another kind of product is produced by the addition of lime or serpentine (a hydrated magnesium silicate) to superphosphate so that the water-soluble P_2O_5 is reverted. Such a mixture must, of necessity, contain less P_2O_5 than the superphosphate, but the object is to prevent the water-soluble P_2O_5 from reverting in the soil to highly-insoluble compounds like iron and aluminium phosphates. This fixation of phosphate is common in certain types of soil especially when they have become seriously acid through neglect of liming. Experimental work on such reverted superphosphates has been in progress for a few years, but it is not possible yet to generalize on their value.

Amongst other products is di-calcium phosphate which is obtained by treating rock phosphate with hydrochloric acid. The acid is a by-product in the manufacture of potassium sulphate from potassium chloride, and the crude di-calcium phosphate is produced as a finely-divided white powder containing about 40 per cent. P_2O_5 soluble in ammonium citrate.

Another fertilizer produced on a limited scale is sulfurophosphate where the rock phosphate is treated with lime and sulphur.

Basic Slag

The composition of slag and its value as a fertilizer are determined by the process of manufacturing steel, of which it is a by-product. Many iron ores contain phosphorus;

much of the ore found in the Midlands of England, and in Lorraine, for example, is relatively rich in phosphorus. As a result of smelting the ores in the blast furnace the phosphorus is recovered in the pig iron and must be removed in converting the crude pig into steel. In the Bessemer converter the molten pig iron is purified by blowing air through it, and Thomas and Gilchrist discovered, in 1877, that the phosphorus could be removed if the converter were lined with a basic material like dolomite and received a charge of lime before the pig iron was poured in. The blast of air must then be maintained until the phosphorus is oxidized, which takes place after the oxidation of the silicon, manganese and carbon. Great heat comes from the oxidation of the phosphorus, of which about 2 per cent. must be present, and the P_2O_5 produced combines with the lime to give a basic slag which is poured off. It is allowed to cool and then crushed and ground so that at least 80 per cent. passes a 100-mesh sieve. It is a heavy, greyish black material containing 18 to 20 per cent. P_2O_5, of which at least 80 per cent., and sometimes more than 90 per cent., is soluble in a 2 per cent. solution of citric acid. It generally contains about 45 to 50 per cent. lime; the bulk of this is combined as phosphate and silicate, but free lime may be present to the extent of 5 or 10 per cent. Consequently, slag should not be mixed with ammonium salts. The slag also contains a great variety of oxides including those of iron, magnesium and manganese.

For some time after Thomas and Gilchrist had introduced their process no use was made of the slag, though it was known to contain so much phosphate. Various methods were tried for extracting the phosphate or bringing it into a soluble form, though without any success; but early in the 'eighties it began to be found that the only thing necessary to make the basic slag available as a manure was to grind it to an extremely fine powder. In this country the value of finely-ground basic slag was first brought to light by Wrightson and Munro in 1885, who carried out a series of experiments, on a chalk soil in Wiltshire and a heavy clay in Durham. The results were highly favourable and showed that the slag was comparable with superphosphate as a source of phosphate to the crop, being much superior to the other insoluble phosphates tried. About the same time as Wrightson and Munro's experiments, basic slag, known in Germany as "Thomasmehl," after one of the discoverers, attracted considerable attention there; many experiments were made with it and proved so successful that the consumption reached a million tons in 10 years. Indeed, so much more quickly was basic slag taken up in Germany, that a considerable export trade from Great Britain at once grew up.

In this country the bulk of the steel is now produced by the basic open hearth process. This employs gaseous fuel to heat the pig iron and scrap, and the reaction time is 4 to 7 hours instead of the 15 to 20 minutes in the Bessemer converter; the quality of the steel produced is therefore under better control. The hearth is built of basic refractory and about 10 to 20 per cent. lime is added with the charge, but the amount of phosphorus need not be so high as in the Bessemer converter for it is no longer required for the production of heat. For some grades of steel, fluorspar (calcium fluoride) is added to assist the conversion; this seriously reduces the value of the slag for agricultural purposes because relatively insoluble fluor-apatite is formed. Hence the amount of phosphorus in the slag and its citric-solubility depend upon the composition of the furnace charge and whether or not it is possible to draw off some of the slag before the addition of the fluorspar.

Grades.—Several types of slag have, therefore, been available in this country in recent years. There have been high soluble slags (over 80 per cent. of the total P_2O_5 soluble in 2 per cent. citric acid) from the Bessemer process, containing 16 to 19 per cent. P_2O_5, and from the basic open-hearth process, containing from 8 to 16 per cent. P_2O_5; there have also been open-hearth slags of 8 to 16 per cent. P_2O_5 but of medium or low solubility—sometimes as low as 20 per cent. soluble in citric acid. The easily soluble phosphorus compound is possibly a combination of tri-calcium phosphate and a calcium silicate of the type $Ca_3(PO_4)_2.2CaO.SiO_2$. The silicates are easily hydrolyzed and can neutralize soil acidity just like limestone. Numerous experiments have shown that a high citric-soluble slag has the same liming value as about half its weight of calcium carbonate, whilst a low soluble slag has only about half this value. The solubility of the slag in citric acid is also a good test of its value as a phosphatic fertilizer as well as a guide to its usefulness in reducing soil acidity. The test consists in shaking 5 g. of slag with 500 c.c. 2 per cent. citric acid for 30 minutes and determining the P_2O_5 in solution. The citric-solubility test for basic slag has been officially recognized for many years on the Continent. In this country, the Fertilisers and Feeding Stuffs Act of 1926 recognizes only water-soluble and total-P_2O_5 values ; in the sale of basic slags, a statement of total-P_2O_5 and fineness of grinding is required, but a supplementary warranty on percentage of citric-soluble P_2O_5—determined according to the F. and F.S. Regulations, 1932—has been general in the trade for some years.

The value of the citric acid test had steadily gained recognition as a result of the experimental work carried out since 1921 by the Permanent Committee on Basic Slag, for an excellent correlation was found between the citric-solubility of slags and their effectiveness in the field. The effectiveness of the slags has been measured both by increase in crop yield and by the recovery of phosphorus in the crop. The following statement on the results of a comprehensive set of pot experiments with rye grass, is taken from the 1936 report of the Committee. ". . . in all stages of growth the yield and phosphoric acid contents of the crops followed the amounts of citric-soluble phosphoric acid extremely closely. . . . the citric acid method gave a surprisingly good estimate of the availability of the slag phosphates." In twelve cuts of rye grass in this particular experiment, 50 to 60 per cent. of the citric-soluble-P_2O_5 of all slags was taken up by the crop. Such a recovery is of course much larger than that found under field conditions. The following are the average figures from 12 experiments with swedes carried out in Scotland in 1934–35, the dressings all being at the rate of 1 cwt. total-P_2O_5 per acre.

Citric-Solubility of Slag .	Control	23·6	44·0	53·0	89·3
Yield in Tons per Acre .	9·0	16·6	17·4	19·8	20·8
Percentage Recovery of P_2O_5	...	6·2	7·2	10·6	13·4
Average Figures from 4 succeeding Oat Experiments (straw and grain)					
Yield in Cwt. per Acre .	65·2	65·3	68·0	69·4	72·4

The 1935 oat crop at 4 centres used for the swede experiment in 1934 was grown without a further dressing of phosphate. The soils varied greatly with respect to general fertility and response to phosphate in 1934. The above figures for the oat crop demonstrate that the residual effects of the slags were not large but were more pronounced with the more soluble slags. There was no tendency for the less soluble to overtake the more soluble, a result confirmed in numerous other experiments.

The total recovery of P_2O_5 is probably best seen in the experimental data collected from grass mown for hay for 3 or 4 years after the application of slag. The greatest recoveries were of the order 20 per cent. from the high soluble slags, whilst the recoveries from the low soluble slags amounted to less than 10 per cent. The value of the citric-solubility test has also been demonstrated in live weight increases in grazing experiments. The general conclusion arrived at, therefore, from the mass of experimental evidence collected in recent years, is that the solubility of the phosphate in 2 per cent. citric acid is a reliable guide to the value of the slag as a fertilizer.

The demand for basic slag is usually greater than the supply and many methods of making low grade slags suitable for agricultural use have been practised or, at least, proposed as practicable. The phosphorus content of the slag may be increased by feeding the slag back in the furnace. The citric-solubility of the phosphate may be increased by bringing a jet of water in contact with the molten slag, or by introducing silica sand or soda ash into the molten slag. The last-mentioned method is said to be particularly effective so that its adoption would seem to depend mainly upon the cost of the treatment.

Production.—The total production of basic slag is of course determined by steel requirements and its suitability for agriculture by the process of steel manufacture. As far as statistics are available, they demonstrate that practically all the agricultural basic slag is produced and consumed in Europe and that Germany accounted for about half of it in 1938. In 1952, the production as P_2O_5 in millions of metric tons was :

Germany	0·26
Belgium-Luxemburg	0·24
France	0·26
United Kingdom	0·07
Total in Europe	0·84

In 1952–53 in the O.E.E.C. countries, basic slag accounted for 34 per cent. of the total fertilizer P_2O_5 of 2·2 million tons. Germany and Belgium each used much more slag than superphosphate, whilst the United Kingdom used about half as much P_2O_5 in slag as in superphosphate.

CHAPTER VI

THE FUNCTION AND USE OF PHOSPHATIC FERTILIZERS

Effects of phosphate—Response to phosphate—Availability—Comparison of fertilizers, fertilizer placement, relative costs, special conditions.

BEFORE considering the question of the relative fertilizing value of the different phosphatic manures and their application in practice, it will be necessary to obtain some idea of the function of phosphate in the nutrition of the plant.

Effects of Phosphate.—It is in some way closely bound up with grain formation, being always found in greater proportions in the reproductive parts of the plant than elsewhere.

This ripening action is very clearly seen in the Rothamsted experiments on barley; the plots without phosphate are, as a rule, about a week behind those which receive this fertilizer.

This effect is brought out in the diagram, Fig. 2, which shows the results of certain determinations made upon barley cut at regular intervals during the development of the grain from some of the Rothamsted barley plots in 1904. The

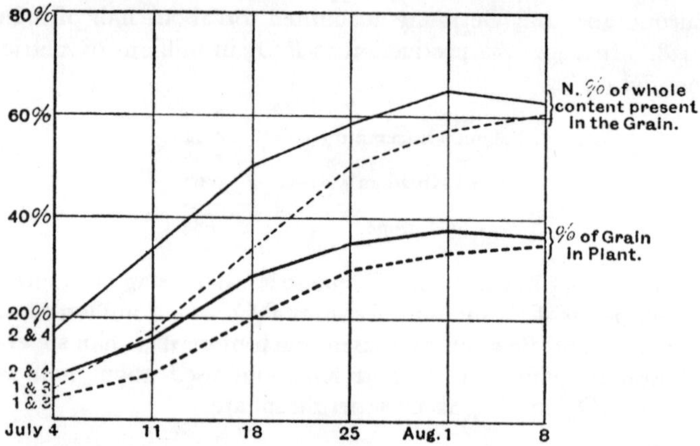

FIG. 2.

Curves showing the effect of Phosphoric Acid in hastening the formation of Grain of Barley, and the Migration of Nitrogen to the Grain. 2 and 4 with Phosphoric Acid. 1 and 3 without Phosphoric Acid.

two lower curves show the rate of the formation of the grain week by week, calculated as percentages of the weight of the whole plant, for the two plots which receive phosphoric acid and for the corresponding plots without, both series being similarly treated as regards nitrogen and potassium. It will be seen that the formation of grain begins earlier where phosphatic fertilizers have been used, and even at the end is more complete. Similarly, the two upper curves show the migration of the nitrogen to the grain, again calculated as percentages of the nitrogen of the whole plant, and, as before, the movement of nitrogen begins at an earlier date, and is more completely carried out when there is plenty of phosphate.

The action of phosphate on the plant is not confined to its ripening effect; it stimulates the early development of the young seedling to a remarkable extent. Farmers are well acquainted with the good start that any crop gets when manured with superphosphate; indeed, it is often used merely to secure a better plant, though with little expectation of otherwise increasing the yield. More than ninety years ago this had been noticed by the late Sir John Lawes; and in one of his earliest papers on "Turnip Culture" in 1847, he writes: "Whether or not superphosphate of lime owes much of its effect to its chemical actions in the soil, it is certainly true that it causes a much enhanced development of the underground collective apparatus of the plant, especially of lateral and fibrous root." For this statement he was vigorously attacked by Liebig, but many experiments have shown that it was the result of sound observation and that, in some way or other, phosphate does stimulate the root development of the young plant. Barley seedlings, for example, grown in water cultures without, or with a minimal amount of, phosphate develop practically no root, whereas when they are nitrogen or potassium starved the root growth is proportional to that of the rest of the plant. Both in the field and in pot experiments the phosphate has a great effect in promoting the formation of adventitious buds, so leading to the tillering of the plant. Hence phosphatic manuring has a valuable effect in establishing the plant, even if the gross yield is not ultimately much enhanced.

Increase in root growth also explains the extraordinary results of quite small dressings of phosphate upon soils in Southern Australia, where a manuring with one cwt. per acre or even less of superphosphate has been found sometimes to increase the yield of cereals by 50 per cent. On analysis the soils are not rich, but they show no such signal deficiency in phosphate as would account for the action of the manure; it seems much more likely that in a semi-arid country where the whole success of the crop depends on the roots getting quickly down to the cooler and moister subsoil, the stimulating action of the phosphate upon the young roots becomes of the greatest value. In this connection it may be noted that two crops which respond well to phosphatic manuring, turnips and oats, are both possessed of shallow roots, confined to a comparatively limited layer of soil; whereas, under ordinary farming conditions, wheat responds very little to phosphate, being a deep-rooted plant. For mangolds, farmers are accustomed to use superphosphate, because they have found by experience it is of great assistance in securing a plant, and there are many soils on which a crop of swedes cannot be grown without a dressing of phosphate.

It has sometimes been stated that phosphate is associated with the assimilation of nitrogen by the plant, and particularly with its migration from the stem or roots into the seed, the opinion being probably founded on the fact that the nucleoproteins, so characteristic of the reproductive parts of plants, contain phosphorus. This opinion is not, however, borne out by the examination of a large number of analyses of barley grain from the Rothamsted plots; when phosphate is deficient the intake of nitrogen is not proportionally reduced: in fact, the grain grown on the plots receiving no phosphate is the richest in nitrogen.

This point may be further elucidated from some experiments upon wheat made at Rothamsted in 1907; a large number of ears of wheat were marked just as they came into flower, in order to secure that all should be as nearly as possible the same age at starting. A number of these heads were gathered every third day, and the grain was extracted and analysed, so as to trace any progressive changes in composition as the grain formed and ripened. Table 16 shows the ratio of nitrogen to phosphorus pentoxide in such grain from three of the Rothamsted plots—from the unmanured Plot 3, where all the elements of nutrition and particularly nitrogen are lacking; from Plot 10, where there is an excess of nitrogen and a great deficiency of phosphate; and from an

TABLE 16.—DEVELOPMENT OF WHEAT GRAIN, 1907.

Date.	Plot 3. Unmanured.		Plot 10. Nitrogen only.		Manuring. Normal	
	Nitrogen per cent. in Dry Grain.	Ratio of Nitrogen to P_2O_5.	Nitrogen per cent. in Dry Grain.	Ratio of Nitrogen to P_2O_5.	Nitrogen per cent. in Dry Grain.	Ratio of Nitrogen to P_2O_5.
July 16 .	2·68	2·15	...	...	...	...
„ 19 .	2·41	2·15	2·61	2·35	...	...
„ 22 .	2·46	2·12	2·45	2·32	...	...
„ 25 .	2·17	2·00	2·13	2·25	...	...
„ 28 .	2·12	2·05	2·10	2·53	...	...
„ 31 .	2·06	2·22	2·11	2·32	2·27	1·91
Aug. 3 .	1·86	1·87	1·92	2·27	2·20	1·88
„ 6 .	1·83	2·04	1·85	2·50	2·05	1·91
„ 9 .	1·80	1·89	1·88	2·45	1·97	1·85
„ 12 .	1·72	1·86	1·83	2·41	1·84	1·82
„ 15 .	1·86	2·28	1·83	2·17	1·85	2·00
„ 18 .	1·79	1·96	1·88	2·29	1·89	1·81
„ 21 .	1·85	2·06	1·90	2·38	1·88	1·82
„ 24 .	1·78	1·99	...	...	2·03	1·91
„ 27 .	...	...	...	...	1·98	1·85
„ 30 .	...	...	...	...	2·05	1·98

adjoining plot under ordinary farming conditions, where phosphate is relatively abundant. It will be seen that on any plot the ratio of nitrogen to phosphorus pentoxide remains fairly constant throughout the whole development of the grain, but that a different ratio exists for each plot. From this we may conclude that the material which the plant on any particular plot moves from its straw and root to form into grain is the same throughout the development of the grain, but that each plant, according to the soil and manure conditions under which it is growing, builds up a type of grain of a composition special to itself. It is true that the nitrogen and phosphate move into the grain *pari passu*, and in that sense the phosphate might be regarded as a carrier of the nitrogen, but then the starch also migrates in an equally constant ratio to the nitrogen compounds, though no such association is claimed between nitrogen and starch. Both actually and relatively the nitrogen is highest in the grain from Plot 10, where the plant is phosphate starved.

On the other hand, it has been shown in U.S.A. that the application of a phosphate fertilizer to a phosphate-deficient soil will usually increase the protein content of alfalfa. The reason is that there is a close relationship between available phosphate and the formation of the nodules so important for the successful growth of leguminous crops. One of the most striking effects of phosphates is to be seen in their encouragement of wild white clover in pastures. A marked improvement of pasture on phosphate-deficient land through phosphate manuring is generally associated with a considerable increase in the percentage of phosphorus in the dry matter of the crop.

When a soil is very deficient in available phosphate there may be a practically complete failure of the crop, a state of affairs which has not been uncommon after the ploughing up of old pastures on acid soils. Where the deficiency is not so pronounced the leaves tend to be dull and bluish green, sometimes with a purple tint; there is a restricted and spindly growth with delayed flowering and premature leaf fall. The most sensitive crops are swedes, turnips and potatoes; hay is normally the least affected.

Response to Phosphate

In their comprehensive analysis of the results of several thousand field experiments in Northern Europe, Crowther and Yates obtained the following average figures for the

Without Dung.		With Dung.
Swedes . .	4·4	2·3 (2·5)
Mangolds . .	2·4	1·0 (1·4)
Sugar Beet . .	0·41	0·66 (0·61)
Potatoes . .	0·84	0·55 (0·61)
Wheat . . .	0·3 (1·0)	...
Barley . . .	0·6 (1·4)	...
Oats . . .	1·5 (2·0)	...

response in Great Britain to 0·5 cwt. P_2O_5, corresponding to about 3 cwt. superphosphate per acre. The figures for roots are in tons per acre, those for cereals in cwt. grain per acre.

The corresponding figures for the general means, weighted according to the number of experiments in each country, are given in brackets. The results are probably typical of a fair cross-section of normal farm soils but, of course, large variations are to be found amongst soils of different type and in areas of different rainfall. In this country the responses are much greater in the wetter regions than in the drier ones. This may be on account of the fact that soils tend to become more deficient in lime as the rainfall increases so that the efficiency of added phosphatic fertilizers decreases; it may also be due to the period of growth being unduly curtailed in the drier areas by the effect of the phosphate in hastening the maturation of the crops. A considerable reduction in response is found when dung is used, although not so much as might be expected from the amount of phosphate in the usual dressing of dung. Sugar beet is exceptional in this respect according to these figures.

In comparing the different crops it will be seen that swedes showed an outstanding response—in absence of dung it amounted to from about 20 to 50 per cent. according to district; mangolds and potatoes showed about half that percentage response whilst sugar beet was less responsive. The average responses of the cereals were not large, but again it is necessary to bear in mind the nature of, and manurial practice for, the preceding crop. The question of the most profitable dressing for a crop is, of course, bound up with current prices, but, by plotting response against increasing dressings of plant food, it is possible to determine from the curve the most efficient dressing for a particular set of conditions. For crops grown without dung in Central and Northern England the average responses in tons per acre of root crops and cwt. per acre of grain were found to be as follows:

P_2O_5 (cwt.).	Swedes.	Potatoes.	Beet.	Mangolds.	Cereals.
0·2	1·8	0·5	0·4	1·0	0·5
0·6	4·0	1·1	0·9	2·2	1·1
1·0	5·0	1·4	1·1	2·8	1·4

It was then calculated that the most profitable dressings at 1940 prices were, in terms of cwt. P_2O_5 per acre: potatoes 1·3, swedes 1·2, beet 0·9, mangolds 0·7 and cereals 0·24, all without dung. In dry areas these figures were reduced by about 0·2 and in wet regions increased by about 0·3. Where dung was used, a reduction of about 0·4 was recommended.

From a survey of fertilizer practice it was found that the average rates of phosphatic dressing applied were, with the exception of the cereals, very much less than those "most profitable dressings." It is necessary to keep in mind, however, that great differences exist from farm to farm and from region to region in the use of fertilizers, and that wartime values were necessarily abnormal.* However, these figures provide a most useful guide to the farmer. The average increase in crop yields from these optimal dressings amounted to swedes 6, mangolds 3, potatoes 1·7, and beet 1·2 tons per acre, and cereals 0·8 cwt. per acre; from these data it is possible to decide which crops are likely to give the best net return from phosphate manuring.

Availability

The phosphatic manures are, in the main, compounds of phosphoric acid and calcium, and several combinations exist and are found in commerce. Only one, the di-hydrogen calcium phosphate, the characteristic constituent of superphosphate, is readily soluble in water ; the others give rise to extremely dilute solutions of phosphate, too dilute, as has been shown by experiment, to nourish a plant properly, with however large a volume of the solution it may be in

* Recalculated for 1954, these dressings become 1·6 for potatoes, 1·4 for swedes, 1·0 for beet and mangolds and 0·4 for cereals. For corresponding yield decreases see p. 248.

contact. Yet, when they are sufficiently finely divided and so incorporated with the soil so as to be in contact with the roots, they are all effective in supplying the plant with phosphate. The carbonic acid excreted by the roots of the plants is the chief agent in producing a solution in the soil water capable of attacking these insoluble phosphates ; acid humus may also assist in rendering them available to the plant.

Since all the commercial phosphates have to enter into solution before they can be utilized by the plant, the question of their relative availability is very important, and various solvents have been proposed for its determination in the laboratory. Generally, for example, basic slag is valued, not on the total amount of phosphoric acid it contains, but on the amount that is soluble in a 2 per cent. solution of citric acid. In U.S.A., a solution of ammonium citrate is used for the valuation of phosphatic fertilizers. It is doubtful whether these solvents, however, really draw a sharp distinction between the different phosphates, which are all soluble up to a certain point, when an equilibrium is established between the phosphate in solution and that remaining undissolved. If the first solution formed is replaced by a fresh portion of the solvent, more phosphate will come into solution; in fact, most of the phosphates can eventually be dissolved by the solvents in question. The following table (17) shows the amount of phosphate extracted by 1 per cent. citric acid from one of the Broadbalk soils manured for fifty years with superphosphate, the extraction being repeated with fresh solvent as soon as one portion had been saturated and then removed:

TABLE 17.—100 g. BROADBALK SOIL (PLOT 7), WITH 1 LITRE 1 PER CENT SOLUTION OF CITRIC ACID.

First Extraction . .	56·1 mg. P_2O_5 dissolved
Second „ . .	22·8 „ „
Third „ . .	8·9 „ „
Fourth „ . .	6·5 „ „
Fifth „ . .	4·4 „ „
Sixth „ . .	4·4 „ „

Very similar results have been obtained when manures are treated in the same manner, and they may be taken to show that a single extraction by any solvent of the kind proposed does not dissolve the whole of a particular compound of phosphoric acid, which may be thereupon reckoned as distinct in kind from the rest of the phosphates left unattacked. This mode of attack with weak solvents is regarded as affording only empirical figures to assist the analyst in forming a judgment of the manure; and the conditions of making the solution, such as time of shaking, temperature, and relative amounts of solvent and substance, must be strictly defined. The only solvent which has any *a priori* justification is a solution of carbon dioxide, such as does the work in the soil, but its concentration is not easy to control; the acids of the cell sap, to resemble which citric acid was taken, have no direct contact with the soil particles.

The kind of information which is yielded by the attack of dilute solvents may be seen in Table 18, which shows some of the results obtained by J. K. S. Dixon when certain phosphates of similar character were shaken with a 2 per cent. solution of citric acid. The results agree in the main with practical

TABLE 18.—RELATIVE SOLUBILITY OF VARIOUS PHOSPHATIC MANURES (Dixon).

Manure.	Nitrogen.	Total P_2O_5.	P_2O_5 dissolved by 2 per cent. Citric Acid Solution as per cent of total P_2O_5.
	Per cent.	Per cent.	
Steamed Bone Meal . .	1·86	29·07	64·29
Steamed Bone Flour . .	1·07	29·14	70·55
English Bone Meal . .	5·17	22·46	56·67
Indian ,, ,, . .	3·35	23·19	52·29
Peruvian Guano . .	1·40	27·28	66·05
,, ,, . .	3·26	21·36	85·95
,, ,, . .	8·11	13·13	95·73

experience and with the field trials which have been made upon these materials; the phosphate of bone meal is less soluble than that of steamed bone flour, and the Indian bones which have been dried and exposed for a long time show a lower solubility than do the fresh English bones. The younger the guano is, as indicated by the increased percentage of nitrogen, the greater is the solubility of its phosphate. But while it would thus be possible by the use of one of these weak acid solvents to arrange the various phosphates in a scale of decreasing solubility, the order of the table would not necessarily represent their relative value in practice.

In considering the action of the various phosphatic manures in the field, the most important factor to be taken into account is the soil, for the relative value of the fertilizers will change entirely with different types of soil. For example, the choice between superphosphate and basic slag as a phosphatic manure must be determined, not by their comparative solubility, but by the nature of the soil and the wetness or dryness of the locality. Numerous experiments have been made to compare these fertilizers, but without resulting in any very general information, just because the question is really settled by those external soil factors which are generally unrecorded. On certain soils one or other of these manures will always give the best results, on other soils their effects may be so much alike that the choice between them can be settled by price alone or by any consideration of convenience that may arise. For example, in one of the Rothamsted experiments one series of plots received superphosphate, another series basic slag, and a third bone meal, in quantities supplying the same phosphate to each, the plots being otherwise treated alike as regards nitrogen and potassium. If we abstract the results which refer to the yields in the year of the application of each manure and reduce them to a common standard each year by taking the yields of the unmanured plot as 100, we obtain the relative figures in Table 19.

It will be seen that the phosphates have produced a greater effect upon swedes and barley than upon the deeper rooting

TABLE 19.—RELATIVE YIELD FROM VARIOUS PHOSPHATES (Rothamsted). UNMANURED = 100.

Crop.	Superphosphate.	Basic Slag.	Bone Meal.
Swedes	120	116	126
Barley	119	121	110
Mangolds	114	105	111
Wheat	106	108	117
Swedes	132	109	131
Means	118	112	119

and more slowly growing mangolds and wheat, but that on the whole the three fertilizers are equally valuable as sources of phosphate on the Rothamsted soil. The soil of the Little Hoos field, in which the experiment was conducted, contains a reasonable amount of lime; it is also fairly heavy so that it retains sufficient moisture to give the phosphates of basic slag and bone meal an opportunity of coming into solution.

Without attempting any detailed review of the numerous experiments with different phosphatic fertilizers, we may yet draw certain general conclusions from them.

On nearly all normal soils, superphosphate is the most effective phosphatic fertilizer when equal amounts of phosphorus added are compared. The exceptions are the light sands and gravels very deficient in lime, peaty soils where the humus is of the sour acid type, and all other soils that have developed an acid reaction. On the peaty soils of the fen country superphosphate is the fertilizer most valued, but there the humus is of the "mild" type, consisting of calcium humate, on which the superphosphate is very effective.

When superphosphate is applied to the soil, the soluble phosphate it contains is rapidly precipitated. How thorough is the precipitation of the phosphoric acid within the soil may be seen from Dyer's examination (Table 20) of the soils from the Broadbalk wheat field, which had been receiving 3½ cwt. per acre of "high-grade super" for fifty years previously.

TABLE 20.—PHOSPHORIC ACID SOLUBLE IN FIVE EXTRACTIONS WITH 1 PER CENT. CITRIC ACID, COMPARED WITH THAT IN MANURE AND CROP (Rothamsted).

	Phosphoric Acid, lb. per acre.			
	Supplied in Manure.	Removed in Crop.	Surplus in Soil.	Dissolved by 1 per cent. Citric Acid.
Broadbalk, Plot 3 .	...	550	- 550	565
,, ,, 5 .	3960	790	3170	3000
,, ,, 7 .	3810	1370	2440	2470
,, ,, 8 .	3810	1520	2290	2055
Hoos, Plot 1 .	...	555	- 555	400
,, ,, 2 .	3390	1200	2190	2315
,, ,, 4 .	3390	1240	2150	2000

He found that though the surface soil to the depth of 9 inches had been enormously enriched in phosphate soluble in 1 per cent. citric acid, the subsoil below had gained practically none, so complete had been the precipitation in the layer stirred by the plough. Again, the drainage waters from these plots show a most trifling amount of phosphate, so that losses by washing out must be negligible. Still more evidence of the retention of phosphate by the soil has been obtained by applying the method of successive extractions with a 1 per cent. solution of citric acid, until the phosphate going into solution has fallen to the low constant figure indicating the solubility, not of the recently added, but of the firmly fixed soil phosphates. About five extractions remove the phosphate down to this point, further extractions remove very little more, and the sum of the phosphate dissolved in these five extractions approximates very closely to the surplus of phosphate supplied as superphosphate over that removed in the crop.

This shows that phosphate supplied as superphosphate may remain in the surface soil, and in a form that is readily soluble in a dilute solution of citric acid. The same result, however, is not obtained if the soil is not well provided with lime; in that case the more insoluble phosphates of iron and aluminium would be formed. It is a fair conclusion to draw from these results that superphosphate and, indeed, all phosphatic manures, may be applied to the land much earlier than is usually the case without fear of their being washed out. Much of the benefit of phosphatic manures is often lost because they are applied when the land has already begun to run short of water. Fine grinding and early application are the two great factors in making phosphatic manures available. But it should also be remembered that only about 20 or 25 per cent. of the phosphate added to a soil is normally recovered in the crops and that the remainder may be fixed in the soil in a form unavailable to the plant.

Comparison of Phosphatic Fertilizers

As already indicated, it is impossible to generalize about the relative values of different types of phosphatic fertilizer without taking into consideration the kind of crop, the type of soil and the climate conditions in question. There are, however, certain obvious principles to be attended to in assessing the probable response to the application of a phosphate. The first effect of phosphate is to encourage root development; hence the promotion of early growth will be obtained if there is a supply of readily available phosphate close to the young seedling. That means that a water-soluble phosphate will be particularly effective on a soil which will not hasten reversion of the phosphate to an unavailable form—to hydroxy apatite on a calcareous soil, for example, or iron and aluminium phosphates on a very acid soil. On acid soils, a phosphate which is insoluble in water is likely to be better, provided there is adequate moisture, and it is in a fine state of division so that a large surface is presented for solution.

Then some crops, especially swedes, seem to be able to make better use of the less soluble forms of phosphate than other crops; potatoes and cereals, on the other hand, generally respond best to the more easily available forms. Apart from the responsiveness of different crops there is the question of the length of the growing season. With potatoes and spring-sown cereals the growing season is relatively short so that a good start is of the utmost importance, and a source of readily-soluble phosphate is highly desirable. The speeding up of the ripening process is also of inestimable value for the lifting of potatoes and the harvesting of grain in late districts. In the case of swedes, a better start may be secured with a soluble fertilizer like superphosphate, but the long growing season often enables a less soluble phosphate to make up the leeway and produce the same final yield if climatic conditions have been such as to bring it steadily into solution.

Fertilizer Placement.—Experience has shown that on reasonably fertile soils, superphosphate has generally been the most effective of the usual phosphatic fertilizers and appreciation of the ideas discussed above has shown how its effectiveness may be enhanced on those soils in which the formation of inert forms of phosphate is liable to take place. It has been customary in this country to broadcast fertilizers in order to secure an even and wide distribution through the soil. In other countries where large areas of cereals are grown, combine drills were introduced to sow the seed and fertilizer at the same time and so save labour and time. Much of the land so treated was seriously deficient in available phosphate and it was discovered that very small dressings commonly gave remarkably good results. The reason obviously was that the bulk of the phosphate, being laid close to the germinating seed, became at once available to the young plant so that little was lost by fixation in the more remote soil particles. Considerable attention has, therefore, been paid in recent years to the development of drills which will place a relatively small amount of fertilizer a short distance from the seed. The object is to lay down a band of the fertilizer near enough to provide the young roots with a rich source of plant food, but not so close as to risk damaging the young plant when nitrogenous or potassic compounds are used. Much investigation is at present going on to decide the best conditions for different crops and many special drills have been devised for the purpose in U.S.A. The value of the combine drill in this country has been fully confirmed by numerous experiments. At normal rates of dressing on phosphate deficient soils, it has been established that superphosphate is twice as efficient when drilled with the seed as when broadcast. On an average, 0·25 cwt. P_2O_5 drilled gives the same increase in yield of grain as 0·5 cwt. P_2O_5 broadcast. Combine-drilling is generally responsible for a quicker establishment of the crop and an earlier ripening; these effects particularly commend themselves to the farmer even if the increase in yield is less than 20 per cent. and so not apparent to the eye. Little advantage is obtained by the placement of nitrogenous or potassic fertilizers, and obviously care must be taken to avoid concentrations that may be harmful to the seed. However, the root growth of most crops is stimulated by the placement of mixed fertilizers near the seed.

Another method of making the fullest use of a water-soluble phosphate is to use it in granular form. In this manner a much smaller surface area comes into contact with the soil so that a smaller proportion of phosphate is liable to be rendered insoluble. Small pockets rich in available plant food are thereby left for the roots of the growing plant. Other advantages of granulation, as previously mentioned, are better storage and easier drilling.

Relative Costs.—In the sale of phosphatic fertilizers in this country the content of phosphoric acid, expressed as phosphoric anhydride P_2O_5, must be stated. Only the water-soluble P_2O_5 is required in the case of superphosphate. For rock phosphate and slag, only the total-P_2O_5 need be stated, whilst for compound fertilizers both water-soluble and insoluble P_2O_5 are required (p. 107). In other words, only the total P_2O_5 and the P_2O_5 soluble in water are recognized under the Fertilisers and Feeding Stuffs Act. No cognizance is taken of intermediate types of phosphates ; no discrimination is made between the P_2O_5 which is soluble in dilute solvents other than water, and the P_2O_5 which is soluble only in strong solvents and practically unavailable to the plant.

Whether these regulations will be adhered to in the future in the face of accumulating evidence that, for many crops and many soils, water-solubility is not a sure guide to response in the field, remains to be seen. Great progress has been made in the manufacture of the various phosphatic compounds described in the last chapter; in many cases their fertilizing value is comparable to that of superphosphate although their contents of water-soluble P_2O_5 may be quite low. Water-soluble P_2O_5 was a valuable check on the quality of superphosphate in its early days of manufacture, but the wisdom of retaining it as the chief index of "available" phosphate is open to criticism.

The prices of phosphatic fertilizers are compared on the cost per unit of P_2O_5. For example, the cost per unit would be 15 shillings in the case of a fertilizer containing 20 per cent. P_2O_5 and priced at 300 shillings per ton. The prices fluctuate somewhat, but in the spring of 1955 they were quoted as follows :

Fertilizer.	Per cent. P_2O_5	Price per ton.	Price per unit.
		£ s. d.	s. d.
Superphosphate . . .	18	13 4 6	14 8
Basic Slag (high soluble) .	16	7 5 0	9 1
,, ,, ...	13	6 8 6	9 10
,, ,, ...	10	5 13 6	11 4
Mineral Phosphate . . .	29	11 0 0	7 7
Triple superphosphate . .	47	35 19 0	15 4

All these fertilizers carry a subsidy, to be recovered by the purchaser ; it amounts to £4 10s. per ton for superphosphate, £2 8s. for 16 per cent. slag, £3 15s. for mineral phosphate, and £11 15s. for triple superphosphate.

It will be noticed that, in the above table, only the P_2O_5 in superphosphate is soluble in water and that, of the slags (citric solubility 85 per cent.), the best grade is the cheapest per unit P_2O_5. Prices were not available for the various types of "reverted" superphosphate.

Special Conditions.—Now the crops which usually receive the largest dressings of phosphate are potatoes, swedes and grassland. All experimental evidence in this country points to superphosphate as being the most effective fertilizer for potatoes, but on acid soils high-grade basic slag and calcined phosphate are often at least as good as superphosphate for swedes and for reseeding grassland, and so, also, are some types of "reverted" superphosphate. On very poor soils in wet areas, ground mineral phosphate is sometimes comparable to superphosphate for swedes or grass. It is obvious, therefore, that considerable economies may be effected by the proper choice of fertilizer and that the solubility of the phosphate in water does not determine its availability in the field.

The particular value of slag has been in its effectiveness in improving the grass on acid clay soils. The effect is due partly to the phosphate which is readily available under these conditions and partly to the neutralizing value of its calcium compounds. This latter property is also important in cultivated soils which are acid, particularly for the swede crop liable to attack by the club-root fungus. Basic slag, even low-grade, is, of course, an expensive material to be used for liming purposes only, but there are occasions when the "lime" value of even 5 or 10 cwt. is of inestimable value in allowing a crop to establish itself on an acid soil. The citric acid test is the best to assess the relative values of different types of slag because the percentage of citric-soluble P_2O_5 decreases rapidly with increase in the amount of fluorine which forms the highly insoluble fluor-apatite. Low soluble slags are probably best used on infertile acid soils in wet areas, where they reduce the acidity and provide a little phosphate.

Bone meal and steamed bone flour are available in relatively small amounts, but are in great demand and their prices per unit of plant food are high. Bone meal is bound to become available very slowly on account of its physical condition and this may be one reason for its popularity, especially amongst market gardeners. But since market garden soils are usually very rich in readily or slowly available phosphate, the better policy would seem to be to add small amounts of easily soluble phosphate (and nitrogen) at intervals during the growing season. Steamed bone flour is finely ground and so has a much better chance to decompose in the soil. It might be expected to react fairly rapidly in an acid soil, but there are too few experimental data to assess its value.

Superphosphate was the first of the manufactured phosphatic fertilizers to be placed on the market and for 100 years has maintained a predominant position. Its chief competitor has been basic slag and both materials have been of the greatest value in agriculture. Since slag is a by-product of the steel industry its production is limited and the agricultural consumption is not likely to increase unless some reasonably cheap process is developed to make the considerable amounts of low-grade slag suitable as fertilizer. Any increased demand for phosphates must, therefore, be met from the large deposits of mineral phosphate in various parts of the world. The raw rock phosphate can be used profitably as a fertilizer only under certain conditions and the bulk of it will continue to be treated in some way to convert it into a compound which will more readily supply phosphate to the plant. Treatment with sulphuric acid to produce superphosphate has been the common method to break down the more or less inert apatites, but strong heating, coupled with the introduction of silica and a basic material, is also effective, while the controlled "reversion" of superphosphate is a successful method of producing valuable fertilizers. Finally, the use of phosphoric acid in the manufacture of such highly concentrated phosphatic compounds as ammonium phosphate, triple superphosphate, and calcium or potassium metaphosphate, is a development full of promise. There is, therefore, reason for complete confidence in the future supply of phosphatic materials suitable for varying circumstances of crop and soil.

CHAPTER VII

THE POTASSIC FERTILIZERS

Nomenclature—Sources of potassium salts, Germany, Alsace, U.S.A., Spain, Poland, Palestine, U.S.S.R., other supplies—Effects of potassium, cereals, roots, legumes, fruits, disease, maturity, yield—Comparison of different potassium compounds.

ALTHOUGH the water cultures already described, coupled with the results of the Rothamsted experiments even in their early years, showed that of the alkali metals found in the plant's ash only potassium was indispensable, for a long time the salts of potassium could not be obtained in quantities or at a price appropriate to agricultural requirements. Almost the only source of potassium was the crude carbonate or "potashes," which was obtained by dissolving the soluble salts found in wood ashes; and though this was to a small extent supplemented by the potassium nitrate or saltpetre obtained from India, and by a certain amount of potassium sulphate obtained from "kelp"—the ashes of seaweed—no widespread use could be made of potassium salts in farming until the opening up of the great Stassfurt deposits in Germany. The fertilizing value of wood ashes had long been known, and in the south-east of England it had been customary for the hopgrowers to organize a regular system of collection of the ashes of their cottagers, who burned little besides wood, but such a supply was only local and early exhausted.

William Ellis, again, writing in 1750, states that "at Long Marston, in Bucks, is a potash kiln, where they make ashes from bean straw for the most part, and sell a vat of them, which contains 32 five-bushel sacks, which dresses one acre for fourteen shillings, to be shovelled out of a cart or wagon, and throwed over grass land in this month (July) or at any time till Candlemass."

In 1861 the output of potassium salts began from Stassfurt and rapidly grew until, in 1900, no less than 1,158,000 tons were being used for agricultural purposes alone.

Nomenclature.—It is necessary to say something at this stage about the nomenclature of these compounds. All potassic fertilizers are expressed on the basis of potassium oxide (K_2O) commonly known as potash. It is an unfortunate convention for potash is not present as such in potassium salts, and is certainly not absorbed by the plant. Potassium is taken in by the plant as the potassium ion which occurs in all solutions of potassium salts. By far the commonest of the potassium salts is the chloride (KCl) but in the fertilizer trade it is called muriate of potash. Muriatic acid and muriates are the old terms for hydrochloric acid and its salts, the chlorides; they are never used now in chemistry, not even to signify that the material in question is chemically impure. Similarly, sulphate of potash and nitrate of potash or even potash nitrate are used for potassium sulphate (K_2SO_4) and potassium nitrate (KNO_3). It would simplify matters considerably, and be much more sensible, to drop the term potash in favour of potassium and give the compounds concerned their proper chemical names. The use of "NPK" with reference to fertilizers containing nitrogen, phosphorus and potassium is already fairly common for the sake of convenience.

Sources of Potassium Salts

The first large deposits to be exploited for agricultural use were to the north of the Harz mountains, near Stassfurt, in Germany, but deposits were located and worked in several other regions. They lie south of the Harz mountains; in Thuringia; in the Fulda valley between the Rhone and Vogelsburg mountains; in the Hanover area; and in the North German plain. In spite of developments in other parts of the world, between 50 and 60 per cent. of the world's supply of potassium salts came from these German deposits in 1953. The other main sources of supply were Alsace and

U.S.A., with smaller amounts from Spain, Palestine, Poland, and U.S.S.R. The world production in 1952–53 was equivalent to about 5·4 million tons of K_2O ; the consumption in this country was nearly 235,000 tons.

Germany.—Common salt had been manufactured for a long time from the natural brine springs at the old town of Stassfurt. A boring made in 1858 proved the presence of rock-salt at a depth of about 1,000 feet below the surface, but above the rock-salt was found a layer of minerals containing potassium and magnesium salts, which were at first regarded as worthless but became the most valuable substances in the mine, because they constituted the largest available source of potassium. The deposits belong to the Permian (magnesian limestone) geological formation and represent the result of the drying-up in a hot climate of a great lake or sea, which retained some connection with the ocean so as to admit of the constant inflow of fresh sea-water. It has been shown that all the salts of sodium, potassium, magnesium, and calcium which occur in these deposits, were formed at different stages in the concentration and drying-up of a solution originally of the composition of sea-water, and the sequence of their deposition has led to the estimation that a period of about 13,000 years was necessary for the formation of the bed. The sequence of deposits varies somewhat from shaft to shaft, especially where the influx of water in the past has effected some rearrangement in the salts; but in the main, after passing through 600 to 800 feet of red sandstone, limestone, etc., a bed of gypsum is first reached underneath which is a bed of very pure, "younger" rock-salt, which in its turn gives place to a bed of anhydrite (anhydrous calcium sulphate) with some gypsum. Below the anhydrite comes a bed of tough impervious salt clay, which has acted as a water-proof layer and prevented the solution of the highly soluble potassium and magnesium salts immediately below; at the top is a layer 50 to 130 feet thick of carnallite, a crude double chloride of potassium and magnesium which was the main source of the manufactured salts. Below the carnallite comes the "kieserite" region, where this mineral, a crude magnesium sulphate, predominates and, below it again, comes a "polyhalite" region, characterized by the prevalence of this complex sulphate of potassium, calcium and magnesium. The polyhalite overlies the "older" rock-salt, 2,000 feet or more in thickness, interspersed with and underlaid by layers of anhydrite, before the limiting bituminous sandstones are reached. It will be noticed that the bottom layer of anhydrite represents the least soluble salt in sea-water; above it comes the sodium chloride in bulk; while at the top are gathered together the magnesium and potassium salts, which would be the last to remain in solution. The processing was extremely simple in principle; the salts rich in potassium, chiefly carnallite, were mixed and brought into solution, from which products of various grades of purity could be obtained by evaporation. In this way were obtained the commercial sulphates and chlorides of potassium, but the substances chiefly used in agriculture were certain crude salts obtained by grinding suitable mixtures of the original material as mined. Of these the best known in this country was kainit, a mineral which, properly speaking, only occurred in some of the mines where water had formerly access, but which commercially represents a mixture of sulphates and chlorides of sodium, potassium, and magnesium, with about 14 per cent. K_2O.

The Stassfurt salts are white or grey or pink (owing to the presence of a little impurity) gritty powders, which dissolve readily and almost entirely, and are, as a rule, more or less deliquescent. Potassium chloride, and particularly magnesium chloride, are very soluble and readily absorb moisture; and as both chlorides and magnesium are present in all but the purest grades of potassium sulphate, the salts used for manurial purposes are always somewhat deliquescent.

With the exhaustion of the true kainit beds and the necessity of increasing the efficiency of the process, hard salt (13 per cent. K_2O) and sylvinite (14 per cent. K_2O minimum) gradually took the place of carnallite (10 per cent. K_2O) as raw material. The production of potassium chloride from sylvinite, which is a mixture of potassium and sodium chlorides, is based upon the fact that whilst the solubility of the sodium salt does not change much with rise in temperature, the solubility of the potassium salt increases considerably. Hence a hot saturated solution of sylvinite will throw out crystals of potassium chloride on cooling. The process of extracting potassium chloride from carnallite or from hard salt, which is a mixture of potassium chloride and magnesium sulphate, is rather more intricate. One of the by-products is kieserite, $MgSO_4.H_2O$, which has been used in the production of potassium sulphate. When kieserite is treated with hot water, it forms the heptahydrate $MgSO_4.7H_2O$ and when this is treated with a solution of potassium chloride

$$2KCl + 2MgSO_4 = K_2SO_4.MgSO_4 + MgCl_2$$

it gives a potassium-magnesium sulphate. A saturated solution at a temperature of 25°C is best for this reaction, which is reversible. The potassium-magnesium sulphate is either used directly as a fertilizer or is treated with more potassium chloride at 30° to 40°C, whereupon potassium sulphate is deposited.

$$K_2SO_4.MgSO_4 + 2KCl = 2K_2SO_4 + MgCl_2$$

The approximate average percentage composition of the potassic compounds generally available in the U.K. in 1955 is as follows :

Name and Grade.	K_2SO_4	KCl	$MgSO_4$	$MgCl_2$	NaCl
Potash Salt 40% . .	...	63	5·5	1·1	21·2
Sulphate of Potash-Magnesia 26% .	50	...	28·0	...	4
Muriate of Potash 50%	...	79	0·4	0·3	14·5
Muriate of Potash 60%	...	95	0·2	0·1	3·0
Sulphate of Potash 48%	89	...	0·7	0·4	...

The grade is the guaranteed minimum of K_2O. Pure potassium chloride contains the equivalent of 63·2 per cent. K_2O ; pure potassium sulphate, 54·1 per cent. The lower grades contain some insoluble matter, 2 to 3 per cent. of gypsum, and considerable amounts of sodium and magnesium salts, and were used largely near the mines, their low contents of potassium not justifying long transport costs. In this country in 1952, 92 per cent. of the total K_2O consumed was in the form of high grade chloride and 2·4 per cent as the sulphate.

Alsace.—The deposits are found at a depth of about 2,000 feet in an area about 12 miles long and 8 miles wide, lying to the north of Mulhouse and west of the Rhine. They consist essentially of the mineral sylvinite (KCl.NaCl) and the crude salts contain the equivalent of about 14 or 20 or 30 per cent. K_2O, and are notable for the complete absence of magnesium sulphate. At first they were simply ground for use as fertilizer, but in recent years an increasingly large proportion has been treated to obtain potassium chloride and so reduce transport costs and meet the demand for high-grade products. The total production in 1953 was equivalent to rather less than a million tons of K_2O.

Potassium chloride is produced from the sylvinite by the process described above, but the French method of producing potassium sulphate is to treat the chloride with sulphuric acid at 350°C in special furnaces. Potassium hydrogen sulphate and hydrochloric acid are formed; the latter is removed and condensed, the former is heated to 800° or 900°C in presence of fresh potassium chloride and is converted into potassium sulphate with the evolution of more hydrochloric acid.

1. $KCl + H_2SO_4 = KHSO_4 + HCl$
2. $KHSO_4 + KCl = K_2SO_4 + HCl$

U.S.A.—Before 1937, U.S.A. was one of the best markets for German salts, but domestic production has increased so rapidly in recent years that by 1954 it practically met the agricultural consumption of about 1·7 million tons K_2O.

There are three main sources of supply. Searles Lake, in the Mojave Desert of California, lies about 1,600 feet above sea-level and is really a partially dried-up lake of some 12 square miles in area. A very complex brine is pumped into evaporators from wells in the centre of the mass of crystals. Many products are obtained by an involved series of crystallizations, including potassium chloride, of 98 per cent. purity, and potassium sulphate, the carbonate, sulphate and borate of sodium (borax), boric acid, bromine, several bromides and lithium salts.

From the Bonneville Basin or Salduro Marsh of the Salt Lake Basin of Utah, a brine in the clay substratum of the salt crust is obtained. It is concentrated by solar heat to yield a mixture of sodium and potassium chlorides which are separated by a flotation process. The crude mixture is ground until the crystals of the two salts are separated from each other and the ground material is then suspended in a saturated brine with a selective filming agent. The suspension is passed continuously through a series of cells into which air is injected in such a way as to form small bubbles. The bubbles carry that salt which has been covered with a film to the surface as a froth and the froth is removed and purified. The process is so efficient that potassium chloride of nearly 97 per cent. purity is recovered.

The most productive source in U.S.A., however, is the Carlsbad region in New Mexico. Deposits of high-grade salts were discovered in 1926 at a depth of about 1,000 feet in the Permian Basin and have been exploited in three areas. Several valuable minerals are present, but only sylvinite and langbeinite have been worked on a commercial scale so far. A high-grade potassium chloride is produced from the sylvinite by a process similar to that employed in the European factories and the sulphate is manufactured as in the French refineries. Potassium sulphate has also been successfully produced from langbeinite ($K_2SO_4.2MgSO_4$). The langbeinite is washed and treated with a solution of potassium chloride which reacts with the magnesium sulphate to give potassium sulphate and magnesium chloride. With careful adjustment of concentration and temperature the potassium sulphate crystallizes out in a comparatively pure state.

Washed langbeinite is also sold and used directly as a "Sulphate of Potash-Magnesia" similar to the German product, and is a useful fertilizer for correcting both potassium and magnesium deficiencies in the soil.

Spain.—The most important deposits are beds of carnallite and sylvinite in the province of Catalonia, and production is equivalent to something of the order of 170,000 tons K_2O per annum.

Poland.—The chief deposits are situated at Kalusz and Stebnik in the foothills of the Carpathian Mountains, and production was increasing rapidly up to 1938, when it was almost the same as the consumption. Both sylvinite and kainit are found and various grades of potassium chloride up to a purity of 96 per cent. were manufactured.

Palestine.—The Dead Sea is about 50 miles long and 9 miles wide; its average level is 1,300 feet below the Mediterranean, and its maximum depth is 1,300 feet. It is the remnant of a much larger lake forming part of the rift between the Lebanon and the Red Sea. It is fed mainly by the Jordan River and springs; there is no outlet but a large evaporation during eight rainless hot months with continual wind. Its concentration increases with depth but the ratios between the various salts are fairly constant as the following approximate figures show.

	Surface.	360 feet.
KCl	10	16
NaCl	71	87
$MgCl_2$	110	169
$CaCl_2$	31	47
$MgBr_2$	4	7
Total in g. per litre	227	327

There is also a small amount of gypsum. The salts are regarded as coming from the Jordan and the total concentration is about 5 times that of the ocean. The Palestine Potash Company began operations on a commercial scale in 1930, and the annual production had reached about 100,000 tons of 80 per cent. potassium chloride. At the north end of the sea, the brine is pumped from a depth of 175 feet to the highest of a series of large salt pans and allowed to evaporate by solar heat. In the first stage of the evaporation about 90 per cent. of the sodium chloride separates, and then a carnallite containing about 20 per cent. KCl and 33 per cent. $MgCl_2$ crystallizes out. A potassium chloride of 80 per cent. purity is then obtained by careful treatment of the carnallite with cold water. Bromine is an important by-product of the works.

U.S.S.R.—Several deposits of sylvinite and carnallite have been discovered in the région of Solikamsk-Berezniky in the Urals, but the full extent of development has not been published.

Other Supplies.

A very large deposit rich in potassium has been located at 3,500 feet in Saskatchewan whilst potassium salts have also been discovered near the coast of Yorkshire in this country.

Vast deposits of volcanic lava containing the equivalent of about 8 per cent. K_2O in the form of leucite have been the basis of a small industry in Italy. A method of extracting potassium salts from sea water has also been operated in Italy. When sea water is concentrated by solar heat, impure sodium chloride is deposited and can be removed from the mother liquor which is known as "bittern" from the bitter taste imparted to it by magnesium salts. On further concentration of the bittern, a mixed salt called mellahite is obtained. It contains 20 per cent. each of sodium and potassium chlorides and 30 per cent. of magnesium sulphate and, after weathering for some months, is ground and washed with sea water to give schoenite ($K_2SO_4.MgSO_4.6H_2O$), a product similar to the German one. A fairly pure potassium sulphate may also be obtained by treating the schoenite with milk of lime and removing the precipitated magnesium hydroxide and calcium sulphate. There is also a Norwegian process in which about 75 per cent. of the potassium of sea water is precipitated by means of lime and dipicrylamine and filtered off; the precipitate is then decomposed by an acid to get back the dipicrylamine and a potassium salt.

Attempts have been made from time to time to obtain potassium fertilizers from certain minerals like orthoclase felspar by grinding or heating with basic substances. Although a certain measure of success has attended these efforts the results have not been profitable and the processes have never been developed on account of the relative cheapness of soluble potassium salts. The dust which collects in the flues of ferro-manganese and pig-iron furnaces may contain soluble potassium compounds in amounts which make it worth while as a fertilizer. The composition of the flue dust is very variable, but there are sometimes consignments containing as much as 15 or even 20 per cent. K_2O; there are also low-grade dusts with only 5 to 10 per cent. K_2O. But the quantity which comes on the market in this country has never exceeded a few thousand tons per annum; to be fit for agricultural purposes the dust should be practically free of cyanides and sulphides which are toxic to plants.

Wood ashes, particularly from small twigs, contain from 10 to 15 per cent. potassium in terms of K_2O according to species, but coal ash contains very little potassium. The ash from the young fronds of bracken contains up to 50 per cent. K_2O—this figure falls considerably towards the end of the growing season—of which more than half is soluble in water, but the expense of cutting and collecting the bracken makes the cost prohibitive. Finally, fresh seaweed contains about 1 or 2 per cent. K_2O and a certain amount of potassium sulphate used to be obtained from seaweed ash or kelp, but the industry is at present of little account although there may possibly be some revival as projects to harvest seaweed for the extraction of alginates are developed.

Reference has already been made in Chapter III to the potassium nitrate extracted from the nitrate beds in Chile. In addition to 15 per cent. N, it contains the equivalent of

10 per cent. K_2O easily soluble in water, and is therefore a most valuable fertilizer and one for which the demand is likely to remain greater than the available supplies.

The Effects of Potassium

To understand the use of potassic fertilizers in the ordinary routine of farming, it is necessary to inquire into the function of potassium in the nutrition of the plant, for the water culture experiments hitherto quoted only demonstrate that it is one of the indispensable elements. Further inquiry goes to show that in some way potassium is an essential part of the mechanism of the process of assimilation; when it is deficient, the manufacture and translocation from the leaf of carbohydrates, like starch and sugar, are greatly reduced, and in practice it is the crops rich in carbohydrate which are most dependent upon a full supply of potassium. It is also probable that potassium is required in the building up of protein bodies for there is a correlation between the amounts of potassium and protein reserves in seeds. Potassium also seems to influence the absorption or transpiration of water so that crops respond best to potassium fertilizers in dry seasons. Various hypotheses have been put forward to account for the action of potassium, but no completely satisfactory explanation is yet available for the relationship between this plant nutrient and the efficiency of the leaf.

The effects of a deficiency of potassium on the plant depend upon the amount of available nitrogen. When the nitrogen supply is low, the leaves are small and rather pale in colour and wither prematurely. Carbon assimilation is, therefore, small, but the proportion of carbohydrate to protein is normal. The plant is usually stunted and both seed and fruit are reduced in size and total weight. When the supply of nitrogen is plentiful, large but relatively inefficient leaves are formed and the proportion of carbohydrate to protein is abnormally low. This leads to a lack of vigour, a susceptibility to attack by fungi and delayed ripening. There is, therefore, a close connection between the effects of nitrogen and potassium and particular care is necessary to avoid potassium deficiency in all crops which are heavily manured with nitrogen. It should also be pointed out, however, that the worst effects of potassium deficiency on certain annual plants are to be seen when both available phosphate and nitrogen are high. The foliage becomes blue-grey in colour and there is leaf scorch and a premature dying down of the plant.

The symptoms of potassium deficiency are generally most apparent in the leaves and begin on the older leaves with a browning of the tips or of the margins (leaf scorch) sometimes accompanied by brown spots near the margins or between the veins. A marginal leaf scorch may be due to other causes but is a typical symptom of potassium deficiency in the case of fruits and broad-leaved crops. Browning of the tips, and brown spots on the underside of the leaves are characteristic developments in the potato. Marginal browning extending inwards between the veins is usual in sugar beet and mangolds while both tip and marginal browning is common in the cereals of which barley seems to be the most susceptible.

Cereals.—The Rothamsted results with barley grown continuously since 1852 are quite striking, because of the large amount of potassium originally in the soil; it is only during the later years that any deficiency has been manifest on the plots that do not receive this fertilizer. Still, as shown in the following table (21), which gives average results for the fourteen years 1889–1902, the use of potassium has increased both the weight per bushel and the weight of the individual grains:

TABLE 21.—ROTHAMSTED BARLEY (14 YEARS, 1889–1902).

Plot.	Manuring.	Weight. per Bushel.	Weight of 100 Grains.
		Lb.	Grams.
1A	Nitrogen only . . .	52·3	4·03
2A	Nitrogen and Phosphoric Acid	52·2	3·86
3A	Nitrogen and Potash . .	53·3	4·14
4A	Nitrogen, Phosphoric Acid, and Potash . . .	53·8	4·21

The maltsters valuation of the grain produced during the above period was highest for plot 4A and lowest for 1A. The result of withholding potassium so long has been to produce a marked depression in yield; the average yields of grain for plots 4A and 2A were respectively 37·9 and 31·4 bushels per acre. Such effects have not been so marked in normal practice, however, on account of the potassium in the farmyard manure used for the root crops in a rotation; in fact, experiments on barley carried out over a series of years on commercial farms have shown that potassium fertilizers have not much effect provided the nitrogen-potassium balance available to the plant is satisfactory. Similarly with wheat, which may show a large response to potassium on a highly deficient soil, but normally does not require more than is required to balance the nitrogen applied.

Roots.—The effect of potassic manures on the production of a carbohydrate, in this case sugar, is most manifest on the mangold crop. If Nos. 4 and 5 (Table 22) of the Rothamsted mangold plots which receive the same supply of nitrogen and phosphate are compared, it will be found that, in a good year, they produce approximately the same weight of leaf; indeed, the similarity would be still closer if the comparison were made when the leaves were in full activity, and not at the end of the growing season. One plot (4), however, receives a dressing of potassium salts, but not the other (5), and plot 4 produces nearly two and a half times the weight

TABLE 22.—EFFECT OF POTASH ON THE PRODUCE OF MANGOLDS AT ROTHAMSTED, 1900.

Plot.	Manure.	Leaf per acre.	Roots per acre.	Sugar per acre.
		Tons.	Tons.	Tons.
5A	Ammonium Salts and Superphosphate . . .	2·95	12·00	0·797
4A	Ammonium Salts, Superphosphate and Potash .	3·25	28·95	2·223

of roots grown upon plot 5. Now the difference in dry weight is almost wholly due to sugar and other carbohydrates, which were manufactured in the leaf and then passed on to the root for storage; yet the two plots possessed practically the same leaf development, working under identical conditions of illumination, carbon dioxide, and water supply. But, in one case, the photo-synthetical process had been limited by the want of potassium; all the machinery was there and the power was in excess, but the machinery was running idle for the lack of one necessary link—in this case the potassium.

The effect of potassium upon the mangold crop is also to be seen upon the plots where dung is also supplied, as shown in Table 23:

TABLE 23.—ROTHAMSTED MANGOLDS (12 YEARS, 1895–1906).

Manuring.	No Potash.	+Phosphates and Potash.
	Tons.	Tons.
Dung only	18·6	19·5
,, and Nitrate of Soda	27·7	26·8
,, and Ammonium Salts	21·8	25·9
,, and Rape Cake	24·9	28·6
Dung, Rape Cake, and Ammonium Salts	24·2	29·9

Here it will be seen that potassium increased the crop in every case, except where nitrate of soda had been used as the nitrogenous cross dressing, in which case the sodium acted as an effective nutrient for at least part of the potassium.

In the earlier years of this experiment only phosphates had been added to the dung and nitrogenous manures, but had produced no increase; we may conclude, therefore, that in the series quoted it was the potassium alone that had been active. This result is the more striking in that dung itself contains a considerable amount of potassium, yet the use of 14 tons of dung per acre year after year, beginning in

1856, has still been unable to supply the mangold crop with all the potassium it needed.

The part played by sodium is illustrated by the results obtained in 1900 (a good year); Table 24 shows the yield of roots and leaves, and the potash and soda removed from the soil by the crop. On plot 5N, without any potassium, but where sodium nitrate is used, the yield of roots and leaves is almost as great as that obtained on 6N where potassium salts are also added, and is more than double that given by the corresponding plot without potassium, 5A, but which receives its nitrogen as ammonium salts. The amount of potash taken from the soil by the crop on Plot 5N is 92·7 lb. against 59·6 lb. on 5A, the increase being small compared with that on 6A and 6N. Since the sum of the two alkalis, potash and soda together, in the crops on 5N, 6A, and 6N is nearly the same,

TABLE 24.—POTASH AND SODA CONTAINED IN MANGOLDS. ROTHAMSTED, 1900.

Plot.		Roots.	Leaves.	Potash.	Soda.
		Tons.	Tons.	Lb.	Lb.
5A	Ammonium Salts and Superphosphate .	12·00	2·95	59·6	56·9
6A	Ammonium Salts, Superphosphate, and Potash	28·20	3·60	306·6	67·0
5N	Nitrate of Soda and Superphosphate .	28·35	3·85	92·7	251·6
6N	Nitrate of Soda, Superphosphate, and Potash	29·65	3·60	220·9	160·6

it may be concluded that on 5N the plant was able to use sodium in place of potassium. The element sodium is an important nutrient in other cases and will be dealt with more fully in Chapter X.

The effect of potassium on sugar beet is to increase slightly the percentage of sugar in the root and increase the total yield of sugar. Nitrogen produces a relatively larger increase of crop but an actual decrease in the percentage of sugar. Hence the great importance of using both nitrogen and potassium fertilizers to obtain the best effect from each. This interaction of potassium and nitrogen is also apparent with potatoes, but is complicated by a further reaction between nitrogen and phosphorus, particularly at lower levels of available potassium. In practice, potatoes respond very profitably to substantial dressings of potassic fertilizers, a fact which is recognized by the high proportion of potassium in the more common "potato compound fertilizers."

Legumes.—Next to their effect upon carbohydrate-making crops, the most striking action of potassic manures is their value in promoting the growth of clover and all leguminous crops. The function of potassium here may be still that of promoting assimilation, because the bacteria which fix nitrogen in the nodules on the roots of the leguminous plants must be supplied with carbohydrate by the plant in order to obtain, by its oxidation, the energy requisite for the fixation of nitrogen. It is, of course, necessary to remember that the legume bacteria live independently of their host and are consequently dependent upon other nutrients besides potassium; an adequate supply of calcium, for example, is particularly important and reference has already been made to the need for phosphate. Nodule production increases when a potassium deficiency in the soil is made good, but excess potassium in relation to calcium may be responsible for a reduced nodulation and nitrogen fixation. However, it is usually found in practice that, if the calcium status of the soil is satisfactory, the growth of legumes is very much promoted by a free supply of potassium, and this is very manifest upon sands and gravelly soils which are often poor in potassium.

This effect may be very strikingly seen when the fertilizer is applied to grassland carrying a mixed herbage, for the potassium encourages the leguminous plants until the aspect of the vegetation may be entirely changed. On the Rothamsted grassland which is mown for hay every year, one plot gets a complete mineral manure—phosphates and sulphates of potassium, sodium and magnesium; the adjoining plot receives the same phosphate, magnesium and sodium, but no potassium, while a third plot gets the phosphates alone. The Table 25 shows the comparative yield and the composition of the herbage by weight:

TABLE 25.—ROTHAMSTED HAY CROP, WITHOUT AND WITH POTASH.

Plot.	Manuring.	Dry Hay.		Composition of Herbage in 1902.		
		1856 to 1902.	1893 to 1902.	Grasses.	Leguminous Plants.	Weeds.
		Cwt.	Cwt.	Per cent.	Per cent.	Per cent.
7	Complete Mineral Manure . .	38·8	36·5	20·3	55·3	24·4
8	Do. without Potash	28·1	21·6	28·8	22·1	49·1
4	Superphosphate only	23·3	17·8	54·4	15·4	30·2
3	Unmanured . .	21·9	15·9	34·3	7·5	58·2

On the plots receiving a mineral manure including potassium half the vegetation consists of leguminous plants, but in the absence of potassium the proportion is only 22 per cent. and 15 per cent. It should be noticed that the large amount of phosphate received by these two latter plots does not result in any great stimulus to the leguminous plants, which constitute 7·5 per cent. of the herbage of the unmanured plot. Where nitrogen is applied and potassium omitted, no leguminous plants are to be found.

Another very striking effect of potassium manuring is also very obvious on these grass plots. On the potassium-starved plots, the grasses fail to a large extent to develop any seed, and the heads are soft and barren, presumably because of the deficiency in carbohydrate formation. For the same cause the straw, not only of the grasses, but also on the similarly manured wheat and barley plots, is always weak and brittle when potassium is deficient. The plants of the potassium-starved plots at Rothamsted are also characterized by those

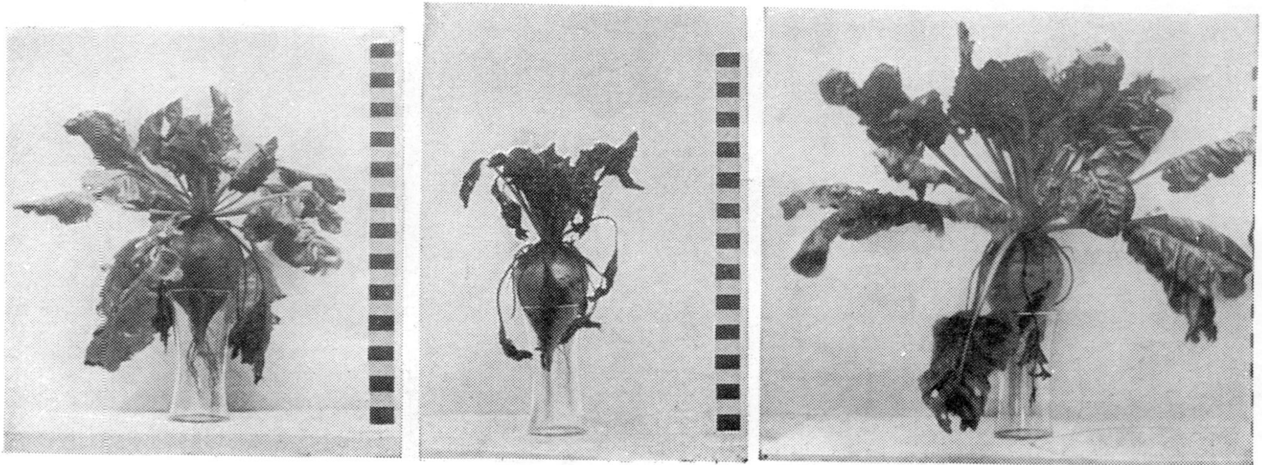

A B C

FIG. 3.—EFFECT OF EXCESS OF NITROGEN, WITH AND WITHOUT POTASH, ON THE LEAVES OF MANGOLDS

A. Normal Manuring
B. Excess of Nitrogen
C. Excess of Nitrogen, but Potash also given

symptoms of potassium deficiency already described. The grass has a dull colour, partly due to a browning along the stems, and partly because the tops of the grass blades show a great tendency to die off for an inch or two and leave a brown withered end.

Fruits.—Troubles due to potassium deficiency have been particularly marked in certain fruit-growing regions and, as in the case of other crops, have often been shown to be due to an unsatisfactory balance between potassium and nitrogen. Leaf scorch is a characteristic and striking feature in apples and pears and some of the soft fruits, and is associated with relatively low figures for the potassium content of the foliage.

Disease.—There is abundant experimental evidence to show that potassium makes the plant more resistant to the attack of fungoid diseases. It has already been explained how susceptible the use of nitrogenous manures renders the mangolds on certain of the Rothamsted plots to the attack of a leaf spot fungus—*Uromyces betae*. The attack is, however, much less severe on the plots receiving an abundant supply of potassium; there the plant remains healthy even though the nitrogen is in excess. The photograph, Fig. 3, shows two typical roots taken in 1902 from plots with and without potassium, both receiving the same large dressing of nitrogenous manures.

In the same way, the wheat on the potassium-starved plots is always subject to rust, even in a good season when very little is to be seen on the other plots normally manured. The grass on potassium-starved plots is also attacked by various fungi; hence it may be taken as a general rule that crops which do not receive their full supply of potassium will be correspondingly susceptible to disease.

Maturity.—It is not possible to say whether the above effect is due to any specific alteration in the composition of the cell contents or to a general lack of vigour, but the latter is probable, because an excess of potassium tends to prolong the vegetative growth of the plant and to delay maturity. Plants receiving potassium are always a little the greener, especially late in the season. This is not always an advantage on heavy soils but is often a great benefit on very dry soils or on soils very deficient in potassium. That potassium tends to prolong growth may also be inferred from the fact that its effect upon the yield is always most pronounced in dry seasons. The long-term experiments with wheat and barley at Rothamsted have demonstrated convincingly that the response to potassic fertilizer is greater in bad seasons of low yield than in good seasons. Apart from its value in raising the yields (compared with the potassium deficient plots), potassium reduced the fluctuations caused by the different distributions of rainfall from year to year, and in the case of wheat, maintained a high ratio of grain to straw.

Yield.—Figures for the average responses of crops to nitrogen and phosphorus have been quoted from the summary of experimental results by Crowther and Yates and their figures for potassium are equally valuable as a guide. The average increases in the yields of root crops, in presence of dung, were of the same order in Great Britain as over Northern Europe as a whole; for 0·5 cwt. K_2O per acre, they amounted to mangolds 1·6, swedes 1·0, potatoes 0·55, and sugar beet 0·25 tons per acre. Without dung, the responses were two or three times as large. In the case of cereals, the same amount of potassium, in absence of dung, increased the yield by 0·7 cwt. per acre of grain.

The most profitable dressings at 1954 prices and the average crop responses to these dressings were calculated to be, in absence of dung:

	Cwt. K_2O per acre.	Response per acre.
Potatoes . . .	1·8	1·8 tons
Mangolds . .	1·4	4·8 tons
Swedes . . .	1·3	2·9 ,,
Sugar Beet . .	1·0	0·7 ,,
Cereals . . .	0·4	0·5 cwt.

The outstanding value of potassium for the potato crop is quite apparent.

Potassium is not readily leached from the soil and the loss in the drainage water is much less than the loss of calcium. It exists in the soil partly in an easily exchangeable form and partly in a form unavailable to plants and there is some reason to believe that one form may pass into the other. Less than half of the potassium added to the soil in the form of fertilizer is recovered in the crops.

Comparison of Different Potassium Compounds

It is not possible to make a strict comparison of the various potassium fertilizers on the bases of their contents of potassium because several other elements which play a part in plant nutrition are involved. As the table giving the composition of the more common products shows, the lower grade of potash salt contains large amounts of sodium and magnesium; the former is a useful nutrient for some crops, especially sugar beet and mangolds, whilst magnesium is a constituent of chlorophyll and therefore an essential element in plant growth. The value of those fertilizers, consequently, may not be due entirely to the potassium which they contain. The question of the amounts of chloride and sulphate is also important for chlorides are easily absorbed by the roots and so affect the concentration and composition of the sap, sometimes deleteriously, whilst sulphur is essential in the formation of many plant constituents and is present in substantial quantity in the brassicæ and leguminous plants.

With respect to price, potassic fertilizers are sold on the basis of their potassium content in terms of K_2O, and the unit of K_2O has generally been cheapest in the muriate. Wherever the sulphate is manufactured from potassium chloride it will be the more expensive salt, and, wherever freight charges are high, the high-grade chloride is likely to be cheaper than the low-grade mixed salts.

Particular attention has been given to responses of the potato crop to the different fertilizers. In most cases, unfortunately, only total yields have been measured, although considerable variations in the percentage of dry matter in tubers is possible as a result of diverse treatments. There are, however, many results which include yields of dry matter and they show that the sulphate has in general been more efficient than those salts containing chlorides. All the evidence points to the fact that magnesium exerts little effect, except, of course, in such soils as may be deficient in this element, and that sodium has no appreciable influence on the potato crop. The differences between the different fertilizers must be attributed, therefore, to the amounts of chloride which they contain. An examination of the available data collected in this country has been made by Cowie whose chief conclusions are summarized below.

Comparisons were made between potassium sulphate, potassium chloride and Kainit (mainly a sodium and potassium chloride mixture) all at the same rate of 0·5 cwt. K_2O per acre. Compared with the response to potassium sulphate, containing no chloride, the yields from the potassium chloride and the kainit showed these reductions in cwt. per acre.

	Potassium Chloride.	Kainit.
Cwt. Cl per acre . .	0·5	2·0
Without Dung—108 trials .	2·0	11·6
With Dung—78 trials .	1·2	7·2

The decreases in response due to potassium chloride are not significant (p. 320), but those in response to kainit are highly significant; the mean depressions in yield are equivalent to about 4·4 cwt. potatoes per 1 cwt. of chlorine. The chlorine also reduced the percentage dry matter of the tubers; 1·4 per cent. in the case of potassium chloride and 2·7 per cent. in the case of kainit, where all the fertilizers were used at the rates of 1 cwt. K_2O per acre in absence of dung or 0·75 cwt. K_2O per acre in presence of dung. The kainit was also responsible for a significant decrease in the percentage of starch in the dry matter and, under certain conditions, to a deterioration in the cooking quality of the tubers. The reasons for the harmful effect of the chloride ion are not yet clear; dung seems to reduce the bad effect as far as crop yield is concerned, but does not exert much influence in the reduction in dry matter; in fact, large dressings of dung also reduce the dry matter content of potatoes.

These different potassic fertilizers have also been compared on fruit and market garden crops. The chloride-containing substances may raise the chlorine content of soft fruits to dangerous levels—especially in the case of red currants—and produce a leaf scorch similar to that caused by potassium deficiency. A chlorosis on indoor tomatoes and on lettuce may also be produced by potassium chloride. All the evidence suggests that, where large amounts of potassium are required in intensive horticultural practice, the use of potassium chloride is attended with some risk of injury to the plants and that the sulphate is the safer form of fertilizer to use.

The examples that have been given to illustrate the specific action of potassium must, of course, be used with some caution as a guide to the manuring of crops under ordinary conditions of farming. Some are extreme cases where the exhaustion of the available potassium in the soil has become very pronounced through the continuous cropping with the help of manures containing all the elements of fertility except potassium. Except on special soils or with the specially potassium-loving crops, it is unusual to find serious crop failure in this country from deficiency of potassium, so well is it conserved by the ordinary systems of cropping.

Potassium is most likely to be deficient on the lighter soils, and the ill-effects are intensified by the dryness of these soils. Potassium deficiency is especially marked on certain chalk soils of the southern counties where even cereals may fail without fertilizer. Otherwise, potassic manures will rarely be found remunerative for cereal crops; for mangolds and potatoes, and to a less extent for turnips, they are necessary, while grassland can hardly be maintained in a satisfactory condition without potassium at regular invervals.

CHAPTER VIII

FARMYARD MANURE

Food, animal, litter—Decomposition, urea, nitrification, putrefaction, humification, gases, product—Losses, preservatives—Composting—Composition of farmyard manure—Management—Value as a fertilizer, nitrogen, phosphorus and potassium, other constituents—Effect on soil texture—Crop response—Cost of farmyard manure.

FARMYARD manure, foldyard manure, yard manure, and dung are all terms employed in various parts of the country for the same more or less decomposed mixture of the excreta of domestic animals with the straw or other litter that is used in the yards or stalls to absorb the liquid portions and keep the animal clean. Probably it would be more correct to retain dung as a name for the excreta alone, and farmyard manure for the product that leaves the yards, but it is impossible in practice to observe any such distinction. It follows, from its origin, that the composition of farmyard manure must be far from constant, varying with the nature of the animal, the kind and amount of food it receives, the proportion between excreta and litter, the nature of the litter, and the extent and character of the decomposition which has taken place in the manure itself. The composition of the excreta being the largest of these factors, it will be necessary first of all to trace the effect of the process of digestion on the various manurial substances in the food—compounds of nitrogen, phosphorus, and potassium.

Food.—Animals that are not increasing in weight, such as working horses or full-grown cattle simply being maintained in store condition, excrete the same amount of nitrogen, phosphorus, and potassium as they receive in a liquid or solid form, the carbohydrates and fat of the food being mostly got rid of as gases. But the fate of the manurial constituents varies according as they are present in the food as digestible or indigestible compounds; for example, part of the proteins of the food withstand the action of the digestive process, and are excreted unchanged along with microbial residues in the *fæces*, but a considerable proportion is broken down into soluble nitrogenous compounds which pass into the blood and eventually are excreted in the *urine*. Similarly, for the phosphorus and the potassium in the food; that which is excreted in the urine is in some simple combination, whilst whatever resists digestion passes out unchanged in the solid excreta. Hence the great difference in the manurial value of the two portions of the excreta; the compounds in the urine—urea, uric acid, soluble phosphates, and potassium salts—are either ready for the nutrition of plants or require but slight further changes to become so; whereas, in the solid dung the materials have several stages of decomposition to go through before they can reach the plant, and having already shown themselves able to resist the attack of the animal's digestive agents they are correspondingly resistant to the ordinary decay processes in the soil. The proportion of digestible to indigestible constituents of a food varies with the nature and even with the mechanical condition of the material, also with the kind and age of the animal; roughly speaking, the richer the food the greater the proportion that is digestible—e.g. decorticated cotton cake contains 7 per cent. of nitrogen, of which 87 per cent. is digestible and finds its way into the urine, while hay contains about 1·5 per cent. of nitrogen, of which only 50 to 60 per cent. is digestible.

Animal.—When the animal consuming the food is growing or fattening or yielding milk, a certain proportion of the manurial constituents in the food is retained, the proportion varying with the nature both of the food and the animal. Cows in milk and young growing animals take the greatest toll from their foods, animals in the later stages of fattening the least.

If, for example, 100 lb. of linseed cake be fed to milch cows or oxen nearly fat, the manurial constituents contained in the cake will be distributed in each case as shown in Table 26.

TABLE 26.—NITROGEN RETAINED AND DIGESTED.

	In 100 lb. Cake.	Fattening Oxen.			Milch Cows.		
		In Meat.	In Urine.	In Fæces.	In Milk.	In Urine.	In Fæces.
Nitrogen . .	4·75	0·21	3·88	0·66	1·32	2·75	0·66
Phosphoric Acid	2·0	0·14	0·09	1·77	0·5	0·07	1·43
Potash . .	1·4	0·02	1·10	0·28	0·14	1·05	0·21

It is thus impossible to state the composition of the excreta of the various farm animals except within certain wide limits, owing to the variations induced by the food and the age of the animal. Table 27 shows certain average results which will serve to characterize the different animals.

TABLE 27.—PERCENTAGE COMPOSITION OF URINE AND EXCRETA.

Animal.	Excreta.	Water.	Nitrogen.	Phosphoric Acid.	Potash.
Horse . .	solid	75·0	0·56	0·35	0·10
	liquid	90·0	1·52	trace	0·92
Cow . .	solid	86·0	0·44	0·12	0·04
	liquid	91·5	1·05	trace	1·36
Sheep . .	solid	57·6	0·72	0·44	...
	liquid	86·5	1·31	0·01	...
Pigs . .	solid	76·0	0·48	0·58	0·36
	liquid	97·6	0·50	0·14	0·70

It will be seen that the urine of sheep and horses is much more concentrated than that of cattle and pigs; similarly, the solid excreta of the two former are also the drier. It is this greater dryness and richness which causes the gardener to describe horse manure as "hotter" than that produced by either cows or pigs; bacterial changes take place in it much more rapidly, a greater amount of ammonia is produced, and the rise of temperature is more pronounced.

Litter.—The next factor which enters into the composition of the manure is the nature of the litter on which the animals are placed; from time to time, especially among small holders, various materials, such as bracken fern, hop bine, leaves, even manufacturing refuse like spent tan and sawdust, are used; but on a large scale only straw and, to a less extent, peat moss litter, are employed. The litter has a twofold function: it absorbs the urine and provides both organic matter and nitrogen for the resulting manure (Table 28). The cereal straws contain about 0·5 per cent. N, 0·2 per cent. P_2O_5 and 1·0 per cent. K_2O, the variations in composition between individual samples of any one kind of straw being as great as the variation between average samples of wheat, oat, and barley straw. Speaking generally, straw grown in the north

TABLE 28.—PERCENTAGE COMPOSITION OF LITTER.

	Water.	Organic Matter.	Ash.	Nitrogen.	P_2O_5.	K_2O.
1. Wheat Straw (Wet Season)	17·8	76·2	6·0	0·38	0·19	0·77
2. Wheat Straw (Dry Season)	15·6	78·9	5·5	0·21	0·17	1·00
3. Oat Straw	16·5	77·9	5·6	0·4	0·28	0·97
4. Barley Straw	20·0	74·6	5·4	0·27	0·18	0·45
5. Bracken	13·6	81·7	4·7	2·0	0·20	0·5
6. Hop Bine	18·7	77·3	4·0	0·28	0·07	0·10
7. Peat Moss	31·8	47·6	20·6	0·83	0·10	0·17

of England and Scotland is richer than straw grown in the south and east of England, because the vegetative growth has been more prolonged and the loss of food materials from the straw has not been quite so thorough. Straw will absorb from two to three times its weight of water, but the variation in absorbing power between different samples of the same kind of straw may be greater than that between different kinds of straw. In practice wheat straw is the most highly esteemed, as cleaner and wearing better under the feet of the animals than any other kind of straw. Oat straw comes next, and is often almost as good as wheat straw; barley straw is least liked, as it is often brittle and dusty.

Peat moss litter is partially decomposed vegetable matter, being derived from the upper layers of a peat bog where the material still retains a good deal of its original structure; this brown, spongy, fibrous mass is sometimes almost entirely organic matter. It can absorb up to about ten times its own weight of water. Peat moss is also remarkable for its power of absorbing ammonia even from the atmosphere, so that a stable littered with peat moss will remain sweet for a comparatively long time. Table 29 shows the result of an experiment in which two similar stables carrying the same stock were littered—the one with straw, the other with peat moss, and the amount of ammonia in the air was determined every day. As will be seen, the peat moss proved a much more efficient absorber of the ammonia produced than the straw. The peat moss itself usually contains a higher proportion of nitrogen than straw does, hence the manure it makes appears to be correspondingly richer, and this difference is often increased by its longer retention in the stalls. But the peat moss itself is very slow to decay so that it is doubtful whether its extra content of nitrogen is of much value; direct experiments, however, are lacking to compare the relative value of manure made from the same amount of feeding stuffs with peat moss and straw respectively. Peat moss

TABLE 29.—AMMONIA IN STABLE PER MILLION OF AIR.

Litter.	1st day.	2nd day.	3rd day.	4th day.	5th day.	6th day.	17th day.
Straw . .	·0012	·0028	·0045	·0081	·0153	·0168	...
Peat Moss . .	0	0	0	0	trace	·001	·017

manure is always short, and is less easy to handle in consequence, but it requires no making and can be applied straight from the yards even to the lightest of soils.

Decomposition

However the farmyard manure has been made, it thus starts with a mixture of excrement, urine, and litter, which become more or less consolidated and mixed together by the trampling of the animals. Other changes, however, intervene very rapidly, and these in the main are brought about by bacteria, which for convenience may be divided into two groups, one acting on the cellulose and other carbon compounds of the straw that make up the bulk of the manure, and the other acting on the nitrogenous compounds.

Urea.—Among the more important of the organisms dealing with nitrogenous material are those which attack the urea in the urine and give rise to ammonium carbonate which very readily dissociates into free ammonia and carbonic acid—both gases, and therefore capable of escaping into the atmosphere.

$$CO(NH_2)_2 + 2H_2O = (NH_4)_2CO_3 = 2NH_3 + CO_2 + H_2O$$

This change into ammonium carbonate is an extremely rapid one; in the liquid draining from a yard or a manure heap, or even in the liquid manure tank, little or no urea can be detected, so complete has been the change to ammonia. As long as the liquid containing the ammonium carbonate is protected from evaporation, no loss of nitrogen will result, but the greater the surface exposed to the air and the higher the temperature, the greater will be the amount of ammonia passing off in a gaseous condition. Thus thin films of urine on the floors, walls, or even on the surface of loose straw, easily lose nitrogen; the smell of a stable arises in this way and is clear evidence of the escape of ammonia. As will be brought out more clearly later, this volatilization of ammonia causes most of the loss of nitrogen that takes place in making dung.

Nitrification.—The ammonium carbonate is itself subject to change and even to loss by other actions than evaporation. There are present in the manure heap various bacteria which can convert ammonia into nitrate and then free nitrogen gas and water. The process called denitrification is the change by which nitrates are reduced to nitrogen gas. Rapid oxidation of the ammonia is brought about by a loose condition of the manure or by turning it, and is also favoured by the presence of soluble carbohydrates.

Putrefaction.—Another group of bacteria which are extremely abundant in fresh fæces are the so-called putrefactive bacteria which break down the proteins into foul-smelling compounds such as amines and mercaptans. Some of these bacteria, like *B. coli communis*, are abundant in the large intestine of herbivorous animals, and of course continue their work in the excreta after ejection. Their function is to convert the insoluble nitrogenous bodies into soluble bodies akin to ammonia and therefore more nearly utilizable by the plant.

Thus, with a certain amount of loss as free nitrogen, the trend of the bacterial actions taking place in the fresh farmyard manure is to break down the complex insoluble compounds of nitrogen to more and more simple ones. At the same time, there is always a reverse change going on; as the bacteria themselves multiply, they seize upon the active soluble forms of nitrogen and convert them into insoluble proteins in their body tissues. Which action is predominant will depend on the stage that has been reached in the dung-making process—i.e. on the supply of carbohydrate, air, water, and other variable factors—but, after the first rapid production of ammonium compounds, the longer the dung is stored the more the ammonia returns to a protein form.

Humification.—So far we have been considering only changes in the nitrogenous material of the excreta and the litter, but the most characteristic change in dung-making is the destruction of the straw and its conversion into dark brown "humus," which in the end retains none of the structure of the original straw. There are a number of organisms to be found commonly in the air and dust which readily attack the carbohydrate material in straw and, in the presence of oxygen, burn it up completely into carbon dioxide and water. For example, an aerobic organism was isolated from soil by Hutchinson which breaks down cellulose with the production of some colouring matter, organic acids like butyric, and a mucilage. Such organisms, however, do not play a major part in manure-making, because oxygen soon gets excluded from the mass; the work is taken up instead by other bacteria capable of working in the absence of oxygen. They rapidly attack carbohydrates like cellulose, and give rise to carbon dioxide, marsh gas, hydrogen, certain fatty acids, of which butyric is the chief, and the indefinite brown acid substance known as "humus." Humus seems to be derived principally from lignin, which is a plant component very resistant to bacterial attack, and protein.

Gases.—Table 30 shows the gases extracted from a fresh dunghill by Dehérain during one of his experiments at

TABLE 30.—COMPOSITION OF GASES IN DUNGHILL (Dehérain).

Date, 1899.	Height of Dunghill.	Point at which Samples were taken.	Temperature.	Carbon Dioxide.	Oxygen.	Marsh Gas.	Hydrogen.	Nitrogen.
	Metres.		°C					
Aug. 22	2·00	middle	52	54·3	0·0	7·8	23·5	14·4
" 23	2·00	middle	52	58·0	...	14·2	11·8	16·0
" 24	2·30	top	71	50·0	...	17·4	3·1	29·5
		middle	67	68·0	...	23·9	7·4	0·7
		bottom	63	49·0	...	40·8	3·9	6·0
" 26	2·30	bottom	60	51·0	...	46·6	2·4	0·0
" 30	2·50	top	60	7·2	7·0	0·0	...	85·8
		middle	65	14·5	4·7	1·3	...	79·5
		bottom	60	50·8	0·0	49·2	...	0·0
Sept. 20	2·50	top	66	42·7	1·1	52·4	...	9·8
		middle	65	49·5	...	48·3	...	2·2
		bottom	52	47·8	...	51·2	...	1·0
Oct. 4	2·50	top	55	54·0	0·5	43·0	...	2·0
		middle	65	42·7	...	56·1	...	0·0
		bottom	40	48·3	...	51·7	...	0·0

Grignon. When the first sample was taken, the dungheap was still in process of formation, and was in too dry a condition. At this stage, hydrogen and carbon dioxide were the most prominent gases. On that day the liquid manure was pumped up over the whole mass, and fermentation became more active, as seen by the very high temperatures reached on the 24th, when the formation of hydrogen had diminished, while that of marsh gas had increased greatly. The analyses on 30th August show the result of having again let the heap get dry; the top and middle were full of air, as may be seen from the large proportions of nitrogen and the presence of some oxygen; the percentage of carbon dioxide had also become so low that losses of ammonia would take place by volatilization, especially as the temperature was high. The later analyses, taken when the heap was well consolidated and kept moist, show that a steady anaerobic fermentation of carbohydrates into equal volumes of carbon dioxide and marsh gas was then going on, while the evolution of hydrogen had stopped.

From these and the other analyses executed by Dehérain, it may be learnt that the main anaerobic fermentation which takes place when the straw and other materials are fresh, is that which gives rise to hydrogen and carbon dioxide; if the heap gets too dry and air penetrates, an aerobic fermentation begins, which gives rise to carbon dioxide only; but at the same time the proportion of this gas falls to such an extent because of its dilution with the air, that ammonia can be lost by volatilization. By consolidating the heap and pumping the liquid over it afresh, the anaerobic fermentation rapidly sets in again and the proportion of carbon dioxide is restored, thus checking the dissociation and volatilization of the ammonium carbonate. After the first outburst of fermentation, the evolution of hydrogen ceases and the marsh-gas fermentation takes its place.

A considerable proportion, amounting to one-quarter or more, of the dry matter of the original dung is lost during this process of humification, by the conversion of carbohydrates into carbon dioxide, marsh gas or hydrogen, and water. The various acids which are also produced are neutralized by the liquid part of the manure, which is alkaline from the presence of ammonium and potassium carbonates; the dark brown liquid to be seen draining from a dunghill is a solution of the humus in this alkaline liquid.

Product.—The changes going on during the making and storage of farmyard manure are thus exceedingly complex; it is in the early stages that the bacterial actions are most rapid, and they fall chiefly upon the carbohydrates and soluble nitrogenous compounds like urea. At this time the greatest losses of nitrogen take place both by volatilization of ammonia and by evolution of nitrogen gas, and so active is the oxidation that the temperature of the mass rises continually. If the rate of oxidation be promoted by occasionally turning over the mass, as in preparing a hot bed or a mushroom heap, the rise in temperature is much increased; at the same time the losses of nitrogen rise rapidly, and the ammonium carbonate disappears more quickly. What the gardener calls "taking the fire" out of the manure, means so reducing the free ammonia that the material is no longer injurious to a plant's roots, though it still remains rich in nitrogen and organic matter capable of further decay. As soon as the first violent reactions are over, expecially after the mass has become consolidated by trampling and the oxygen in the entangled air has been used up, the rate of change slows down considerably; it now consists mainly in the attack of the anaerobic organisms upon the carbohydrate material. The long strawy dung begins to change to "short" or rotten manure, and this change may continue slowly for years, until all trace of structure is entirely gone and only a brown pulp is left. During this second change, but little loss is experienced by the nitrogenous compounds if the mass is kept tightly pressed and moist enough to exclude air. As the manure gets older and shorter it becomes richer in nitrogen; this apparent increase is, however, simply due to the loss of non-nitrogenous carbon compounds, with an increase in the nitrogen to carbon ratio of the residue. But though there is no loss in nitrogen in these later stages, the more active compounds, such as ammonia and the easily decomposable amides, become converted by bacterial action into protein compounds which take longer to reach the plant when the manure finally gets in the soil.

Thus, during the making and storage of farmyard manure there are many bacterial actions at work, and it will depend on such external conditions as the supply of air and water which class of action predominates at any given time. Putrefactive bacteria are resolving proteins into simpler compounds of nitrogen and ultimately into ammonia: nitrifying bacteria produce nitrate from which denitrifying bacteria set free nitrogen gas; meantime the bacteria engaged in the destruction of cellulose and the formation of humus are always building proteins or bodies akin to them out of the previously produced amides and ammonia.

One other change sometimes takes place when the manure is allowed to get too loose and dry—instead of bacteria, fungi begin to develop very rapidly until the whole mass becomes permeated with the mycelium. The masses of manure begin to look white and dusty, a condition which the practical man describes as "fire fanged." It is generally agreed that such manure is seriously deteriorated, but no analyses are available.

Losses

With these general facts in mind it will be possible to interpret the experiments which have been made to ascertain what part of the fertilizing materials contained in foods consumed by animals is recovered in the dung and what losses occur during the making and storage of farmyard manure. In the first place, it can be shown that there is no loss of nitrogen in the gaseous form due to the animal; the nitrogen contained in the urine and fæces is equal to the nitrogen in the food, less whatever may have been retained by the animal in its bodily increase. Numerous feeding experiments demonstrate this point; the following example from Kellner's researches may be taken as an illustration.

An ox was fed on a daily ration of 2 kg. of gluten meal, 2 kg. of starch meal, 4 kg. of dried sugar-beet slices, 5 kg. of hay, and 1 kg. of chaff, containing in all 388·8 g. of nitrogen. About 18·5 kg. of dung was excreted containing 15·36 per cent. of dry matter and 100·9 g. of nitrogen, and about 13 kg. of urine containing 2·03 per cent. of nitrogen, equal to 265·5 g. of nitrogen. The ox was putting on weight, and retained from the food 714·5 g. of carbon and 22·5 g. of nitrogen. Thus of the nitrogen supplied 68·2 per cent. was excreted in the urine, 25·4 per cent. in the fæces, and about 6 per cent. was retained by the animal. To attain such a result, however, it is necessary to collect the urine and fæces as they are voided, and to preserve them or analyse them before any fermentation and evaporation of ammonia can take place.

Assuming the animal itself to cause no loss of nitrogen other than that retained in the increased live weight, a number of experiments have been made to ascertain the losses in making farmyard manure under ordinary working conditions.

For example, Maercker and Schneidewind, at Leuchstadt, in 1896-97 tied up twenty-four three-year-old steers from 16th June to 29th October, 1896—136 days—during which their average increase of live weight was 306 lb. The food consisted of lucerne hay, chaff, barley straw, dried sugar beet pulp, decorticated cotton cake, and bran, and they were littered on wheat straw.

Twelve of the beasts were tied up in a deep, carefully cemented box or pit, from which no losses by drainage could take place, and the dung was not disturbed but kept trampled down until the end of the trial. The second twelve were fed in an ordinary stall, and the dung and litter were removed every other day to one or other of two heaps in the yard alternately, one of these being covered by a roof, and the other open to the weather. At the end of the feeding experiment the three lots of dung were carefully sampled and analysed with the results set out in Table 31 following.

TABLE 31.—LOSS OF NITROGEN IN MAKING AND STORING DUNG (Maercker and Schneidewind).

No.	Conditions under which Manure was made.	Nitrogen supplied.		Nitrogen recovered.		Nitrogen of Food.			Active Nitrogen recovered.	
						Recovered.				
		In Food.	In Litter.	In Meat.	In Dung.	In Meat.	In Dung.	Lost.	Calculated.	Found.
	FIRST EXPERIMENT.									
		Lb.	Lb.	Lb.	Lb.	Per cent.	Per cent.	Per cent.	Per cent.	Per cent.
1	Deep box . .	1033	68·0	64·3	900	6·2	80·6	13·2	70	28·5
2	Covered Heap .	1033	68·0	68·0	652	6·6	56·5	36·9	70	18·0
3	Open heap . .	1033	68·0	68·0	646	6·6	56·0	37·4	70	13·0
	SECOND EXPERIMENT.									
4	Deep box . .	922	52·1	54·6	797	6·0	80·8	13·2	...	...
5	Deep box, dung lying one month untrampled . .	953	65·4	45·9	634	4·8	59·7	35·5	...	...

In a second experiment, fourteen steers were fed in the deep pit from 6th November, 1896, to 21st February, 1897, when the dung made was cleared out, sampled, and analysed. The experiment was then resumed until 21st May, after which the dung was left in the box for another month, until 17th to 18th June, without any beasts to keep it trodden down, the weather being meantime very hot. The results appear under items 4 and 5 in Table 31. It will be seen from this that the loss of nitrogen was much greater during the second series, which only differed from the first in the fact that the dung lay without trampling for a month during the summer.

Taking these results as a whole, it is seen that, even with the most careful management, the loss in making the dung amounts to 13 per cent. of the total nitrogen supplied in the food, in addition to 6 per cent. or so which the animals retain. This loss increases with great rapidity if the conditions are less favourable; the minimum is only attained if the dung is kept trampled beneath the animals in a deep box, for if it is left to itself for a time, or if it is made in a shallow stall and thrown out daily into a heap, as is often the practice, the loss rises to between 30 and 40 per cent.

In connection with the first-mentioned experiment, Maercker and Schneidewind made determinations of the state in which the nitrogen exists in the dung, whether it was soluble and therefore active, or insoluble and comparatively inactive. From the known digestibility of the foods consumed, it was possible to calculate what proportion of the nitrogen in each food left the body in a digested condition as urea and kindred bodies dissolved in the urine, and what proportion consisted of undigested and insoluble compounds in the fæces. Maercker and Schneidewind found not only that the loss had fallen upon the active nitrogen—i.e. that urea had been transformed into ammonium carbonate and volatilized, or broken up with loss of free nitrogen—but also that some of the active nitrogen had been converted into an insoluble form, as though the bacteria swarming in the dung had seized upon the active nitrogen and converted it into the

TABLE 32.—LOSS OF NITROGEN IN MAKING MANURE (Müntz et Girard).

Experiment.		Nitrogen.				For 100 Nitrogen in Food.		
						Recovered.		
		In Food and Litter.	In Meat or Milk.	In Dung.	Lost.	In Meat or Milk.	In Dung.	Lost.
		Kg.	Kg.	Kg.	Kg.	Per cent.	Per cent.	Per cent.
1	Horses in stalls . .	52·4	...	39·9	12·6	...	71·3	28·7
2	2 Cows in boxes . .	14·1	2·8	7·5	3·9	20·0	52·7	27·2
3	8 ,, . .	79·5	10·7	41·7	27·0	14·4	49·3	36·3
4	8 ,, . .	96·6	17·7	50·1	28·8	19·7	48·3	31·9
5	8 Cows and 2 Beasts in boxes . . .	98·8	15·8	49·4	33·5	16·6	48·2	35·2
6	25 Fattening Sheep under cover . . .	14·7	0·9	6·5	7·3	6·2	43·6	50·2
7	25 ,, ,,	23·1	1·4	9·0	12·7	6·0	38·7	55·3
8	20 ,, ,,	14·4	1·4	6·4	6·6	9·9	44·2	45·9
9	20 ,, ,,	14·6	0·2	8·0	6·3	1·7	54·5	43·8
10	20 ,, ,,	10·5	0·6	5·3	4·6	5·5	50·2	44·3
11	Fattening Lambs . .	98·0	10·0	35·4	52·5	10·6	34·1	55·3

No. 1. Littered on straw, dung removed every day.
No. 2. No litter, urine drained off, dung removed several times a day.
Nos. 3, 4, and 5. On litter, dung removed once or twice a week.
Nos. 6 to 11. On litter, dung only removed at end of experiment.

insoluble material of their own substance. Of course, this withdrawal of nitrogen from the active into the insoluble form still further reduces the immediate value of the dung as a whole.

In France, experiments were carried out on the same question by MM. Müntz and Girard, with omnibus horses, cows, and sheep. They showed that with horses and milking cows, where the manure was removed every day, the loss of nitrogen amounted to from 30 to 35 per cent. of the total nitrogen contained in the food. With sheep the losses were still higher. Liberal littering and immediate treading of the excreta into it by the animal greatly reduced this loss. The results are given in Table 32. It is apparent that, on the whole, the proportion of the nitrogen recovered in the manure is about one-half of that supplied in the food.

Further experiments with sheep show that the loss was greatest with no litter, and could be reduced by using an excess, or particularly by using peat moss or earth.

		Loss per cent.			Loss per cent.
Cows	No Litter	59·0	Horses	On Straw	58·0
	Litter	50·2		,, Peat Moss	44·1
	Litter	44·2	Sheep	On Straw	50·2
	Abundant Litter	40·8		,, Earth	25·7

Experiments of the same kind have also been carried out on the farm of the Royal Agricultural Society at Woburn for some years, and the results obtained in 1899, 1900, and 1901, are given in Table 33. The animals were fed in deep boxes with cemented bottoms and sides, and the dung was not removed unti the feeding experiment had concluded; it was then weighed and samples taken for analysis. The manure was then, in the early winter, made up in a heap in the open on ground beaten down hard, and covered thoroughly with earth. No liquid appeared to drain away, and in the spring the heap was again weighed and sampled before application to the land for the root crop. Here, again, the loss of nitrogen in making the dung under the best conditions varied

TABLE 33.—LOSS OF NITROGEN IN MAKING AND STORING FARMYARD MANURE PRODUCED BY CONSUMPTION OF FOODS BY FATTENING BULLOCKS, 1899, 1900, AND 1901 (Woburn).

	For Continuous Wheat Experiments.						For Continuous Barley Experiments.					
	Manure as removed from boxes.			Manure as applied (after storing) to Land.			Manure as removed from Boxes.			Manure as applied (after storing) to Land.		
	1899.	1900.	1901.	1899.	1900.	1901.	1899.	1900.	1901.	1899.	1900.	1901.
	Lb.	Lb.	Lb.	Lb.	Lb.	Lb.	Lb.	Lb.	Lb.	Lb.	Lb.	Lb.
Total Nitrogen estimated to be in Manure, as calculated from Foods consumed after deducting live-weight increase (Lawes and Gilbert) .	34·23	32·39	29·87	34·23	32·39	29·87	34·23	32·39	29·87	34·23	32·39	29·87
Total Nitrogen actually found in Manure . .	29·81	26·47	24·96	21·30	20·92	19·84	32·07	26·66	24·67	22·53	22·78	20·39
Loss of Nitrogen .	4·42	5·92	4·91	12·92	11·47	10·03	2·16	5·73	5·20	11·70	9·61	9·48
Percentage loss of Total Nitrogen .	12·90	18·28	16·44	37·70	35·50	33·60	6·30	17·69	17·41	34·18	29·70	31·80

from 13 to 18 per cent., while the making into a heap and storage brought up the loss to 33–37 per cent.

Wood, at Cambridge, also estimated the losses involved during the making and storage of farmyard manure. In his experiments four heifers were tied up and fed, one pair on mangolds, hay, and straw alone, the other pair on the same foods with the addition of decorticated cotton cake. The feeding went on for 84 days in boxes with well-rammed clay floors, the dung was not disturbed but was kept trampled down by the animals; this is taken as the period of "making" the dung, and at the end samples were drawn by cutting out sections. The dung was now left without moving for six months, May to November, and again sampled as it was taken out—this constitutes the storage period. Table 34 shows the fate of 100 lb. of dry matter and nitrogen respectively fed to the animals.

In an experiment made by Russell and Goodwin at the Wye Agricultural College, the beasts were fed upon roots, hay, and linseed cake, a comparison being made between

TABLE 34.—LOSS OF NITROGEN IN MAKING MANURE (Wood).

	Retained by Animal.	Lost.		Recovered in Dung.	
		During Making.	During Storage.	When Made.	After Storage.
DRY MATTER.					
Roots and Hay only	2·6	38·8	16·2	58·6	42·4
Roots and Hay with Cake . .	5·0	35·0	18·6	60·0	41·4
NITROGEN.					
Roots and Hay only	8·0	16·8	10·6	75·2	64·6
Roots and Hay with Cake . .	9·0	12·5	26·9	78·5	51·6

linseed cake poor and rich in oil respectively. The feeding lasted for twelve weeks and the litter was composed of a bottom layer of peat moss, to which straw was added at the rate of 28 lb. per week. Table 35 shows the results obtained.

TABLE 35.—LOSS OF NITROGEN IN MAKING MANURE (Russell).

	Nitrogen supplied.			Nitrogen recovered.			Nitrogen lost.	
	Digestible.	In-digestible.	In Litter.	In Meat.	Ammonia, etc.	Slowly available.	Actual.	Per cent.
	Lb.	Lb.	Lb.	Lb.	Lb.	Lb.	Lb.	
1. Cake poor in Oil .	43·83	11·77	4·35	3·07	28·6	20·9	7·38	14·9
2. Cake rich in Oil . .	33·29	10·19	4·35	2·04	22·1	16·9	6·79	14·4

The dung was sampled immediately the experiment was over, while the manure was still tight under the feet of the animals; the experiment also took place during the winter months, yet the loss still amounted to nearly 15 per cent. of the total nitrogen. It is noteworthy that all the experiments quoted show practically this same loss of 15 per cent. for the first stage of the dung-making process under the best conditions. Not only does the loss fall upon the active compounds of nitrogen, but a still further amount is converted into more slowly acting bodies. For example, in Experiment 1 there were 43·83 lb. of digestible nitrogen fed, of which the animal only retained 3·07; the remainder, 40·76 lb., was excreted as urea, but only 28·6 lb. of ammoniacal and amide nitrogen were found in the dung, so that besides the loss of 7·38 lb. another 4·78 lb. had been transformed into proteins and other insoluble compounds.

It will be seen that in all cases the losses fall most heavily on the rich dung made by animals receiving concentrated foods; they also fall almost entirely on the most valuable part of the manure—the urea and ammonia compounds arising from the digestible portions of the food.

Further experiments of Russell and Richards at Rothamsted confirm these results. They go to show that a loss of about 15 per cent. of the nitrogen is unavoidable while the manure is being made under the beasts. In practice the dung is best left under the beasts, well compacted, until it can be taken straight out on to the land. Further losses always occur in storage under ordinary working conditions. If the manure can be preserved under strictly anaerobic conditions some 17 per cent. of the dry matter will be converted into gas but there will be no loss of nitrogen. Indeed, in a silo the whole of the straw and other solids can be fermented away, leaving only a liquid sludge which still contains the nitrogen that was originally put in. But when the manure is drawn out and made into the usual heap there will be a loss of both dry matter and nitrogen which will increase with the looseness and exposure of the heap. Even under good conditions of compactness the loss of dry matter in three months may amount to 30 per cent. and of nitrogen to 25 per cent. With loose heaps and with exposure to wind and weather these losses will run up rapidly. Nitrates are formed in the outside of the heap whenever it becomes dry; the best storage is attained by keeping the heap moist and as nearly anaerobic as possible. When, as with dairy cows, the manure is thrown out every day the losses are at a maximum unless it can be put into an undrained pit and well pressed down. The direction of general change depends upon such conditions as the aeration, moisture, and temperature of the mass, and different reactions may take place simultaneously in different parts of the same dunghill.

It has generally been assumed that these losses can be avoided whenever the food is consumed upon the land as when sheep are folded upon the arable, and milch cows or fattening beasts are receiving cake while at grass. It is, however, extremely doubtful whether in the case of sheep grazing upon arable land there is a satisfactory retention of the nitrogen voided by the sheep, such that it can be recovered in the succeeding crops. Much must depend upon the class of land, the weather and the time of year at which the grazing takes place. The increased yield following folding is not always commensurate with the manure residues presumed to be left in the ground, especially where the feeding has been heavy. Many good farmers are persuaded that folding is one of the best ways of maintaining the fertility of the land, but others are not convinced that they obtain a fair return for the food thus consumed.

Of course, in all these considerations no account has been taken of such preventable losses as those which too often occur through the escape of the liquid portions of the manure in a leaky yard or into the drains, or by the washing of rain through the dung heap. Such losses are very great and fall on the most valuable substances in the manure—the ammonia and soluble potassium compounds which occur in the liquid portion.

Preservatives.—In order to minimize these losses of nitrogen, a number of substances have been suggested, which, when strewn about the cattle stalls and mixed with the fresh dung, would either combine with the ammonia and prevent its volatilization, or by reducing the bacterial actions would hinder its formation. These preservatives fall into two classes: those designed merely to fix the ammonia, and the true antiseptics which will check the production of either ammonia or free nitrogen gas. Of the first class of substances, the oldest proposal was to use gypsum, which would react with the ammonium carbonate and form the non-volatile ammonium sulphate:

$$(NH_4)_2CO_3+CaSO_4,\ 2H_2O = (NH_4)_2SO_4+CaCO_3+2H_2O$$

The drawback to the use of gypsum lies in the large quantities that are required; the reaction represented by the above equation is really a reversible one, so that only part of the ammonium carbonate is transformed into sulphate, the amount being proportional to the excess of gypsum present. Since the gypsum is an insoluble salt, far more than the calculated quantity is required for an efficient fixation of the ammonia. Again, the urine contains nearly all its potassium in the form of potassium carbonate, and this also reacts with gypsum, increasing the quantity that must be used before the ammonia is fixed.

From the above equation, about 11 lb. of gypsum are required for each ton of dung, but at least ten times as much as this would be necessary in practice. Besides the question of cost, another great drawback to the use of gypsum lies in the fact that the calcium sulphate is itself liable to bacterial change; during the storage of the dung it is reduced by anaerobic bacteria to the state of calcium sulphide, which afterwards acts injuriously on plant life when the farmyard manure is applied to the soil.

Another suggestion has been to use kainit, because it is composed of salts of magnesium and potassium which will, to a certain extent, be transformed into carbonates and fix the ammonia as chloride or sulphate. Here, again, the quantity required is very large, though the soluble nature of the kainit enables it to be utilized more thoroughly. But of this class of substances the most effective is superphosphate; it reacts with the ammonium carbonate to form tri-calcium phosphate, ammonium sulphate, water, and carbon dioxide. The same objections, however, apply to superphosphate as to gypsum; uneconomical quantities are required if the fixation of the ammonia is to be complete; it has moreover been found to be harmful to the feet of the animals.

Strong acids are highly effective for they not only convert the ammonium carbonate into a non-volatile salt but they make the manure acid and so reduce urea decomposition. Sulphuric acid has been tried, but it is too corrosive to be used except in very dilute form. There is probably more promise in phosphoric acid in view of recent developments in its production from rock phosphate. It has been found to be more effective than superphosphate in reducing losses of ammonia and it increases the water-soluble phosphate content of the manure.

Peat litter exerts an important nitrogen-saving action. It can absorb about 2 per cent. of its dry weight of ammonia and, since it contains relatively small amounts of easily decomposable carbohydrates, it heats less than straw. If it is an acid peat, the alkalinity of the manure is reduced and so the nitrogen is better preserved.

As to antiseptics proper, soluble fluorides, formalin, copper and zinc salts and even carbon bi-sulphide have been tried, but the saving effected in the nitrogen is never sufficient to pay for the cost of the material and the trouble of applying it. Schneidewind, in the course of his experiments at Leuchstadt, found that the only practical means of reducing the losses of nitrogen, was to place a layer of old well-rotted farmyard manure as a basis for the new manure heap; this had a distinctly beneficial effect and always resulted in smaller losses of nitrogen, probably because of the constant evolution of carbonic acid from the layer of old manure.

Composting

An important application of the changes involved in making farmyard manure lies in the treatment of straw or other vegetable refuse without the intervention of the animal It is not suggested that the process will displace the ordinary method of making manure, but there is much to commend it where a larger amount of straw is available that can be trampled by stock, or where market gardeners can no longer obtain the stable manure on which their soil management was largely based. The micro-organisms responsible for the rotting of cellulose require more nitrogen than is present in straw; they also require a suitable amount of moisture, and the acids formed during the decomposition must be neutralized. Investigations have shown that the amount of nitrogen must be raised to about 1·2 per cent. of the dry material. The best compound to use is calcium cyanamide which provides an adequate amount of basic material to prevent the accumulation of acids. Ammonium sulphate may also be used but ground limestone or chalk is required and the two substances should not be allowed to remain in contact on the surface of the straw otherwise they will react and liberate ammonia. Nitrates are not suitable as a source of nitrogen for, under anaerobic conditions at the bottom of the heap, they are denitrified with loss of free nitrogen.

A ton of straw containing about 0·5 per cent. nitrogen requires about 80 or 85 lb. of calcium cyanamide, or the same weight of ammonium sulphate and 100 lb. of limestone. This gives 1·2 per cent. of nitrogen but, during rotting, about 40 per cent. of the organic matter is lost, the bacteria and fungi assimilate the nitrogen, and the dry matter of the final product contains about 2 per cent. nitrogen. If less nitrogen is added to the straw the rotting proceeds more slowly; if excess is added much of it will remain as ammonia whose fate will depend upon the conditions prevailing. About 800 gal. of water are required to supply enough moisture for 1 ton of straw.

Many practical details in composting remain to be settled, but the following method is satisfactory. A layer of straw about 2 feet deep is sprayed with water and left for a few days. A preliminary fermentation takes place and the straw is now better able to absorb more water. The nitrogen compound is then scattered over the straw and washed in followed by the limestone, after which a second layer of straw is added and the process repeated until the heap is 6 to 8 feet high. The temperature of the mass will rise, and this is a useful means of destroying weed seeds, but excessive heating leads to serious losses of organic matter and should be prevented by compressing the heap to exclude air and by adding more water. The process is complete in about six months when the straw has become brown and lost its cohesion.

Many kinds of plant residues, excepting woody tissues, can be dealt with in this way and some, like short grass cuttings, provide a useful amount of nitrogen. Young hedge trimmings, vegetable waste, stalks of brussels sprouts and various weeds all decompose quite rapidly. There is also no reason why animal residues should not be incorporated. Poultry manure or animal urine or sewage sludge are valuable sources of nitrogen, and this is a better method of dealing with liquid manure than storing it in a tank. There is also reason to believe that the addition of a small quantity of well-rotted farmyard manure is a useful starter in the humification of the straw. Straw has also been employed as a filter bed for purifying crude sewage but it is doubtful if this method could be used on anything but a small scale.

Composition

Turning now to the composition of farmyard manure, the average of a large number of analyses at Rothamsted shows that it contains about 76 per cent. of water, 0·64 per cent. N, 0·23 per cent. P_2O_5, and 0·32 per cent. K_2O, or per ton about 15 lb. of nitrogen, 5 lb. of phosphoric acid, and 7 lb. of potash. The composition, however, will vary very greatly, both with the nature and feeding of the animals, and the treatment and storage the manure receives.

The influence of the feeding is well illustrated in a series of analyses of two lots of dung, made in adjoining boxes by bullocks receiving in the one case roots and hay only, and in the other a fattening ration of cake in addition to the roots and hay. The two lots of dung were generally made up into separate mixens out of doors, and sampled a month or two later, when they were carted out to the land; in one case they were sampled as they left the boxes. Table 36 shows the analytical results, not only as regards the total nitrogen, but also that present as ammonium salts, and as amido-compounds easily changing into ammonia.

It will be seen that the cake-fed dung is always considerably richer in nitrogen, the average percentage being 0·73 as against 0·52, a superiority of nearly 40 per cent. Moreover, the extra nitrogen in the cake-fed dung is mostly in the highly available forms, the ammonia, urea, and amido-compounds which represent the digestible nitrogen of the cake. That the superiority of the cake-fed dung as regards the soluble nitrogen compounds is not even more pronounced, is due to

TABLE 36.—PERCENTAGE COMPOSITION OF FARMYARD MANURE MADE AT ROTHAMSTED FROM ROOTS AND HAY ONLY, OR FROM ROOTS AND HAY WITH CAKE.

	Year.	Dry Matter.	Total Nitrogen.	Nitrogen as Ammonia.	Nitrogen as Amides.	Insoluble Nitrogen.	
Roots and Hay only	1904	23·6	0·577	0·046	0·067	0·464	Made into Mixen and stored.
Cake-fed	1904	24·0	0·716	0·079	0·096	0·541	
Roots and Hay only	1905	29·5	0·462	0·040	0·047	0·375	Do.
Cake-fed	1905	31·3	0·698	0·182	0·055	0·461	
Roots and Hay only	1906	22·0	0·466	0·022	0·033	0·411	Do.
Cake-fed	1906	24·3	0·690	0·097	0·049	0·544	
Roots and Hay only	1907	25·3	0·589	0·125	0·053	0·411	Not stored.
Cake-fed	1907	25·5	0·815	0·377	0·033	0·405	

the change back from ammonia into proteins effected by bacteria during storage; in 1907, when the dung was sampled as it left the yard, both lots contained practically the same proportion of insoluble nitrogen, and both possessed an exceptional amount of ammonia, which, however, was three times as much in the cake-fed as in the other manure. These differences in composition are clearly reflected in the crops grown with equal quantities of the two manures, the weights of which are summarized and reduced to a common standard (the yield of the unmanured plots being taken as 100) in Table 37. The crops grown in these trials were swedes, barley, mangolds, and wheat in rotation, and after the two kinds of dung had been applied in a given year, no other manure was used on those plots for the next three years. In the first year the increase in yield produced by the cake-fed dung was 83 per cent., as compared with an increase of 32 per cent. produced by the root and hay dung; in the following

TABLE 37.—CROP RETURNS FROM FARMYARD MANURES.

	Year of Application.	Second Year.	Third Year.
	Mean of 4.	Mean of 4.	Mean of 3.
Unmanured Plot . . .	100	100	100
16 tons per acre Root and Hay Dung	132	131	112
16 tons per acre Cake-fed Dung	183	137	118

year the residue left by the cake-fed dung produced an increase of 37 per cent., as against 31 per cent. from the residue of the other manure; in the third year the increases produced by the residues still remaining were 18 and 12 per cent. respectively. The great difference in the value of the two manures comes in the first year, for though the superiority of the cake-fed dung may still be seen in the second and third year, it is almost covered by the experimental error.

The analyses in Table 38 show the change in composition which results from the storage of farmyard manure; it will

TABLE 38.—COMPOSITION OF FARMYARD MANURE FROM VARIOUS SOURCES.

	Water.	Nitrogen.	Phosphoric Acid.	Potash.
1. Fresh long Straw .	66·17	0·544	0·318	0·673
2. No. 1 after rotting .	75·4	0·597	0·454	0·491
3. Very old and short from a mushroom bed	53·14	0·80	0·63	0·67
4. Fresh (French data)	75·0	0·39	0·18	0·45
5. Rotten (French data)	75·0	0·50	0·26	0·53
6. Very old (French data)	79·0	0·58	0·30	0·50
7. Rothamsted average	76·0	0·64	0·23	0·32
8. Fresh Liquid Manure —Cattle . .	94·0	1·0	0·03	1·0
9. German (300 samples) . .	75·0	0·5	0·24	0·71
10. Swiss (326 samples)	77·0	0·5	0·3	0·6

be seen that old short dung contains a higher proportion of fertilizing constituents (i.e. when reckoned in the dry matter, because the amount of water present at any time is a matter of accident) than fresh dung, if it has been at all properly managed. We have already seen that though considerable losses of nitrogen take place during the rotting down of the manure, the losses of non-nitrogenous organic matter are greater still, so that the manure becomes concentrated in nitrogen and still more so in phosphate and potassium. The active compounds of nitrogen, however, like ammonium carbonate, grow less as the manure ages, since they are constantly being converted into insoluble protein-like bodies making up the bacteria themselves. These, of course, die and decay, giving rise again to soluble nitrogenous compounds, but the tendency is on the whole in the other direction, so that the older the manure the poorer it becomes in ammonia and kindred bodies. Hence old short dung is both slower in its action and less caustic to germinating seedlings or the fresh delicate rootlets of tender plants; it can, in consequence, be used with more safety in the spring in potato drills or immediately beneath the seeds of swedes and mangolds, particularly on a light soil. An analysis of liquid manure is also given; variations due to dilution with rain water, or absorption of its constituents by the litter, will alter the composition considerably. It will be seen, however, that the fertilizing constituents are chiefly nitrogen and potash, both in an active form; hence it forms a very valuable manure for grassland.

Table 39 shows a series of analyses made by B. Dyer of stable manure from London, such as was formerly used in very large quantities by farmers and market gardeners, whose distance from London did not render the freight too great. The most noticeable thing is the very low proportion of nitrogen that remains soluble; the frequency with which the stables were cleaned out in London, the open nature of the heaps, and the many turnings to which the manure was subjected in collection and transit, all resulted in extreme aeration and a rapid fermentation with a corresponding loss

TABLE 39.—COMPOSITION OF LONDON STABLE MANURE (B. Dyer).

	Peat Moss.	Straw.	Mixed Peat Moss and Straw.				
			Fresh.		After Storage.		
			1	2	1	2	3
Water . .	77·8	70·0	76·1	62·0	53·8	61·9	52·9
Organic Matter .	18·0	24·3	19·3	26·4	17·5	22·0	23·0
Nitrogen, soluble	0·51	0·10	0·08	0·08	0·06	0·08	0·10
Nitrogen, insoluble . .	0·37	0·52	0·46	0·62	0·58	0·68	0·79
Phosphoric Acid	0·37	0·48	0·33	0·45	0·49	0·56	0·66
Potash . .	1·02	0·59	0·45	0·58	0·58	0·65	0·80

of ammonia. The last three samples had been stored for eight or nine months on the farm; usually no great care is taken to consolidate such heaps, so that the rotting down process goes on rapidly. In the above cases Dr. Dyer calculated that the loss in organic matter has been about 40 per cent., and in nitrogen from 15 to 20 per cent., during the storage.

Management

From a consideration of the origin of the losses of nitrogen which take place during the making of dung, and of the above analyses, a good deal of guidance can be obtained as to the practical management of farmyard manure. In the first place, since it is clear that the most valuable part of the manure resides in the liquid, far more care should be taken to preserve this than is usually the case. Whether the dung is made in boxes or in yards, there should be sufficient depth to allow the manure to accumulate under the animal for the whole winter if need be, and the floors should be rammed with clay to render them water-tight. Yards, in particular, should be constructed so that the accumulated manure is not above the general ground line outside, in which case there will always be a gradual soaking away of the liquid. On the other hand, yards made thus below the general ground level are apt to flood in heavy rain, so that the excess of liquid containing the soluble part of the manure has to be run off to waste by means of a drain; this can, however, be avoided by cutting drains outside to keep land water from running into the yard, and by seeing that all the surrounding sheds are properly provided with guttering. Only just enough litter should be used to soak up the urine, and in order to prevent the liquid working up to the surface with the trampling, the floor of the yard should run down to a slight hollow, filled at first with something stiff like bean haulm or coarse peat moss, in which the excess of liquid may collect. Above all, the manure should be kept tightly trampled; the greatest loss takes place when the urine falls on a thin layer of loose strawy litter. The yards and boxes should be deep enough to carry the animals through the whole winter, so that they need not be cleaned out except when dung is wanted to go straight on the land.

In many European countries it is customary to collect, store and apply the liquid manure separately. The liquid is drained into tanks where the urea is rapidly converted into ammonium carbonate so that care must be exercised to control the loss of ammonia by evaporation. A tight-fitting cover and a thin layer of crude oil on the surface of the liquid are effective measures to adopt. It is also important to reduce losses in the field by covering the liquid with soil as soon as possible after application. Properly preserved and undiluted liquid manure may contain about 1 per cent. each of nitrogen and potassium, mainly in readily available form and equal in fertilizing value to nitrogenous and potassic fertilizers.

A good deal of attention has also been given, especially in Germany, to the best methods of making and storing farmyard manure. In the hot fermentation method a thin loose layer of manure is allowed to heat rapidly to about 140°F. It is then consolidated and covered by another loose layer which is allowed to heat and so on until a heap of

10 to 20 feet in height is built which is allowed to stand for some months. The principle of the process is that the rise of temperature due to the intense aerobic fermentation destroys weed seeds and all but heat-resistant spore-forming organisms, whilst the slow anaerobic ripening process gives a dark, humified and odourless product. In the cold fermentation process, the fresh manure is tightly packed into a special enclosed compartment so that an anaerobic fermentation takes place with the formation of acids which neutralize and reduce the loss of ammonia; the exclusion of air is such that the temperature does not rise above 90 or 95°F. Obviously a good deal of labour is required for these special procedures and it is not clear whether they are profitable, but the care given to the question illustrates how much farmyard manure is prized.

All experimental evidence points to the serious losses incurred by the exposure of farmyard manure previous to incorporation in the soil. Danish work has demonstrated that, leaving it spread for four days before ploughing may reduce its value to the crop by about 50 per cent. The loss is somewhat reduced by leaving it in small heaps until just before ploughing, but prolonged exposure in this case is responsible for local concentrations in the field of plant nutrients leached out by the rain. It should be observed, however, that less straw is used in Denmark than in this country so that losses of ammonia are likely to be heavier. As far as possible manure made in the spring should be left undisturbed until the autumn; it may then be carted out on to the stubbles and ploughed in where potatoes or roots are to be taken in the following spring. Even on the lightest soils the land will be more benefited thus than if the manure is made up into a mixen and only put on immediately before the roots are grown. Sometimes, of course, a potato grower must have a supply of well-rotted manure to put in the drills immediately before planting; this can often be got from the lower layers of the earliest used boxes or yards, since a mixen should be avoided as much as possible. The principle to keep in mind is that every disturbance of farmyard manure results in loss, and that the shorter the time which elapses between the dropping of the dung and its application to the land, the less this loss of fertilizing material will become.

In considering the value of farmyard manure as a fertilizer one has to keep in mind that it is an essential product of the farm; it constitutes the main source of manure for the land under the conditions of ordinary mixed farming, where inorganic fertilizers will be used to supplement the supply of available plant nutrients. It is only in certain special cases, such as potato or hop growing, where the ordinary course of farming does not supply as much farmyard manure as is wanted, that the question has to be decided whether dung from the towns shall be purchased or whether stock shall be fattened solely with the view of making manure.

Value as a Fertilizer

Nitrogen.—As a fertilizer, the value of farmyard manure lies in the fact that it contains all the essential elements of plant nutrition though the phosphorus is relatively deficient. The nitrogen is present in various forms of combination, varying from the rapidly acting ammonium compounds down to some of the undigested residues which will remain for a long period in the soil before becoming available to the plant. In consequence dung is a lasting manure, which accumulates in the soil to build up what a farmer calls "high condition" —the state of affairs which prevails when the reserves of manure in the soil are steadily and continuously passing into the available condition in sufficient amount for the needs of the crop. But, however marked the farmer's preference is for such lasting manures, the delay in realizing the capital they represent means a certain amount of loss; besides which, some of the constituents of farmyard manure are so slowly acting as to be hardly recoverable during the lifetime of the tenant.

How great the waste may become is seen by comparing the nitrogen supplied to one of the permanent wheat plots at Rothamsted, which receives 14 tons of farmyard manure per acre every year, with the nitrogen stored up in the soil and that removed in the crop. Table 40 shows that only 26 per cent. was recovered in fifty years, and that nearly 57 per cent. has been lost, since it is accounted for neither in the crop nor in the soil at the end of the period. Similarly, the recovery in the mangold roots of the nitrogen added in the continuously dunged plots was only 32 per cent. in comparison with a recovery of 78 per cent. of the much smaller application of nitrogen as sodium nitrate.

TABLE 40.—FATE OF NITROGEN IN FARMYARD MANURE, APPLIED TO WHEAT (Rothamsted).

Plot.	Manuring.	Nitrogen in Soil 9 inches deep, 1893.		Approximate supply of Nitrogen in Manure in 50 years.	Approximate removal of Nitrogen in Crops, 50 years (1844-1893).	Surplus of Nitrogen over Plot 3, unaccounted for in Crop or Soil.
		Per cent.	Pounds per acre.			
				Lb.	Lb.	Lb.
3	Unmanured .	0·0992	2570	...	850	...
2	Farmyard Manure .	0·2270	5150*	10,000	2600	5670

These, however, are extreme cases; on referring to the crops grown with the rich and poor dung (Table 37), where four crops in rotation are grown after each application of farmyard manure, out of 207 lb. of nitrogen supplied as dung made from roots and hay alone 144 lb. were recovered in the three following years, and of 257 lb. supplied as cake-fed dung 158 lb. were similarly recovered.

The extremely lasting character of those nitrogenous compounds in farmyard manure which are not recovered in the first year is illustrated in an exceptional manner in the Rothamsted experiments. On the grassland, for example, one plot received 14 tons of dung per acre per annum for eight years (1856–63) and then was left unmanured. Table 41 shows that it has continued to give a larger crop than the

TABLE 41.—PRODUCE OF HAY PER ACRE, FIRST AND SECOND CROPS, SHOWING RESIDUAL EFFECT OF DUNG (Rothamsted).

Plot.	Manures.	Mean, 8 years (1856–1863).	Season 1864.	Season 1865.	Average of 10 years (1886–1875).	Average of 10 years (1876–1885).	Average of 10 years (1886 1895).	Average of 10 years (1896–1905).
		Lb.	Lb.	Lb.	Lb.	L.b	Lb.	Lb.
2	Farmyard Manure, 8 years (1856–1863), Unmanured since	4804	5392	2848	3726	3748	2791	1943
3	Unmanured continuously . . .	2665	2688	1296	2374	3025	2621	1686
	RELATION TO PRODUCE OF PLOT 3 RECKONED AS 100.							
2	Farmyard Manure, 8 years (1856–1863), Unmanured since	180	201	220	157	124	106	115
3	Unmanured continuously . . .	100	100	100	100	100	100	100

unmanured plot alongside for more than forty years. The table shows that in the first year after the application of farmyard manure had been stopped, the plot with the residues of the previous eight years' manuring gave double the yield of the unmanured plot; in the following year the yield was still double; but from that time its superiority slowly declined, though for the last ten years it still amounted to 15 per cent. The effect is not, of course, due to nitrogen alone.

In considering these results it must be remembered that such a long duration of the residues of farmyard manure would not be perceptible in practice: they only become apparent when the soils are cropped to a state of exhaustion that should never be met with in ordinary farming experience.

Since only a small portion of the nitrogen of farmyard manure is readily available, and if it is the only manure supplied, the crop is often unable to obtain nitrogen rapidly enough in a good season, even though very large quantities are lying dormant in the soil. As an example, we may take the Rothamsted mangold crops for the years 1900 and 1907, when crops considerably above the average were grown, and compare the yields obtained with farmyard manure used alone, with a dressing of inorganic salts containing nitrate of soda and with farmyard manure supplemented by nitrate of soda (Table 42).

TABLE 42.—YIELD OF MANGOLDS AT ROTHAMSTED, 1900 AND 1907. ROOTS ONLY.

Year.	Farmyard Manure + Phosphoric Acid and Potash =200 lb. N.	Nitrate of Soda + Phosphoric Acid and Potash = 86 lb. N.	Farmyard Manure = 200 lb. N. + Nitrate of Soda = 86 lb. N.	Farmyard Manure = 200 lb. N. + Nitrate of Soda + Phosphoric Acid and Potash = 86 lb. N.
	Tons.	Tons.	Tons.	Tons.
1900	28·0	33·1	41·3	41·8
1907	26·5	32·8	41·4	42·1

The farmyard manure, though it contains about 200 lb. of nitrogen, could not provide the rapidly growing mangolds with as much nitrogen as the nitrate of soda containing 86 lb. of nitrogen, since it only grew 27·2 tons of mangolds against 33 tons with nitrate of soda, and this notwithstanding the great accumulation in the soil of the residues of thirty years' previous manuring with dung. Both the plots received the same phosphate and potassium and the crop had by no means reached its maximum, for an addition of nitrate of soda to the dung increased the crop to 41·4 tons. Here, again, only the nitrogen is concerned, because, on a further plot where phosphate and potassium were added to the combination of dung and nitrate of soda, there was but a very slight additional increase of crop.

Phosphorus and Potassium.—The recovery of added nitrogen is much less from farmyard manure than from nitrogenous fertilizers, but the position is different for phosphorus and potassium. The recovery of phosphorus is sometimes higher from farmyard manure than from phosphatic fertilizers and the most likely explanation is that the organic phosphorus compounds are not rendered inert so easily as some inorganic phosphates under certain soil conditions, whilst the extra carbon dioxide produced by the decomposition of the organic matter speeds up the solution of the phosphatic compounds. The recovery of potassium is also practically the same as that from potassic fertilizers. But farmyard manure does more than supply the crop with nitrogen, phosphorus and potassium; it contains a portion of all the other elements, many of them essential for plant growth, present in the original litter and food of the animal, and it improves the physical condition of the soil, thereby giving the plant a better chance to respond to suitable treatment of the soil.

Other Constituents.—Substantial amounts of calcium and magnesium are introduced into the soil with farmyard manure, and the finely divided colloidal humus which is formed, increases the absorption capacity of the soil for bases added in fertilizers. In view of the fairly widespread occurrence of various deficiency diseases it is also worth while bearing in mind that a normal dressing of dung contains a pound or two of manganese and a few ounces each of copper and boron, three of the essential trace elements (Chap. XI).

Some of the nitrogen products of the decomposition of organic residues are derivatives of indole. Many indole compounds are plant growth-promoting substances; in very small concentrations they stimulate root development. The sex hormones of animals are also known to encourage plant growth—at least in culture solution. It has, therefore, been regarded as a distinct possibility that these substances might be responsible for the special merits of farmyard manure claimed by many growers. Experimental work has so far failed, however, to confirm that these compounds play an important part under field conditions (Chap. X).

Effect on Soil Texture

The value of farmyard manure to the land is by no means confined to its fertilizing action; its physical effects upon the texture and water-holding powers of the soil are equally important; indeed, for some crops, and particularly in droughty seasons, these factors count for more than fertilizers towards ensuring a good yield. The farmyard manure, as it rots down in the soil, goes to restore the stock of humus, which is always tending to diminish, and the humus, considered merely from the physical side, contributes largely to the fertility of the soil. It is colloidal and improves the texture of all soils. It gives cohesion and water-retaining power to sands and makes sticky soils more friable and permeable by binding together the finest particles of clay. The improvement in the aeration of a heavy soil thus attained is of the greatest importance in promoting the healthy development of plant roots.

When a piece of old grassland, even on the stiffest of soils, has been ploughed up it is easy to see the beneficial effect of the humus that has been accumulated; after the winter the plough slice will crumble naturally so as to harrow down at once to a mellow seed-bed, whereas a neighbouring piece of the same soil that has long been under arable cultivation may show only a number of harsh intractable clods. The importance of a good seed-bed to the future well-being and ultimate yield of the crop can hardly be exaggerated. It is the basis of all good farming so that even when the fertilizing properties of farmyard manure have been replaced by inorganic compounds, some other means, such as the ploughing-in of green crops, must be resorted to in order to maintain the stock of humus. Of course, the value of humus—and in this respect of farmyard manure—will vary on different soils and with different crops. Cereals, for example, are less affected than root crops which are very dependent on a mellow seed-bed. This may be seen on the Rothamsted plots; the wheat which has now been grown on the same land for nearly a century, comes as well and yields as big crops on the plots receiving only inorganic manures as it does on the plot receiving dung, but on the mangold field the result is different. Where fertilizers containing no organic matter have been supplied, the tilth is bad, and in trying seasons, when drought succeeds heavy rain soon after sowing, the plant obtained is so imperfect as to reduce the yield considerably. If the conditions are favourable to germination and the plant once becomes established, then, as we have previously seen in Table 42, the plot manured with minerals and sodium nitrate may grow a bigger crop than that receiving dung; but this superiority is masked in many seasons by the defective plant resulting from the bad texture of the soil.

The plot receiving rape cake (2,000 lb. every year) is actually better as regards the number of plants it carries than the dunged plot, because the repeated dressings of an organic manure like rape cake supply enough humus to maintain the texture without getting the soil too open—a defect which is now beginning to overtake the plot that has been so continuously treated with large amounts of farmyard manure.

A soil which has been enriched in humus by repeated applications of farmyard manure will resist drought better than one in which the humus is low. The difference is due partly to the greater amount of moisture present in the soil containing humus, and partly to the way it will absorb a large amount of water temporarily during heavy rainfall and then let it work more slowly down into the soil, thus keeping it longer within reach of the crop. Good examples are afforded by the Rothamsted plots; samples of soil were taken from the wheat land on 13th September, 1904; on the previous day 0·262 inch of rain had fallen, but for nine days before there had been little or no rain. The portions of the plots from which the samples were drawn had been fallowed through the summer, so that the drying effect of the crop is eliminated. Samples were also taken from the barley plots on 3rd October of the same year; 0·456 inch of rain had fallen on the 30th September, before which there had been fifteen days of fine weather. Table 43 shows the water in the soil of the unmanured and the continuously dunged plots, calculated as percentages of the fine earth from which the stones had been sifted.

TABLE 43.—PERCENTAGES OF WATER IN ROTHAMSTED SOILS.

Depth.	Broadbalk Wheat.		Hoos Barley.	
	Unmanured.	Dunged.	Unmanured.	Dunged.
Inches.				
0 to 9	16·0	19·3	17·0	20·7
9 ,, 18	19·8	17·0	22·5	17·7
18 ,, 27	23·3	18·4	22·1	18·3

It is seen that in both cases the dunged soil, rich in humus, had retained more of the comparatively recent rainfall near the surface, so that the top soil was moister, while the subsoil was drier. The difference in favour of the surface soil is about 3·5 per cent., which on that soil would amount to about 30 tons per acre, or approximately 0·3 inch of rain. It is thus seen that the surface soil of the dunged plot had retained practically the whole of the preceding rainfall: and the greater dryness of the subsoil is due to the way the soil has kept back the small rainfalls, which have evaporated instead of being passed on to the subsoil, as happens on the unmanured plots. The same fact is illustrated by the behaviour of the drains, which lie below the centre of each of the wheat plots at a depth of 30 inches; below the dunged plot the drain very rarely runs—only after an exceptionally heavy and long-continued fall; whereas the drain below the unmanured plot runs two or three times every winter. Putting aside the greater drying effect of the much larger crop on the dunged plot, the difference is mainly due to the way the surface soil rich in humus absorbs more of the water at first, and then lets the excess percolate so much more slowly that the descending layer of over-saturation, which causes the drain to run, rarely or never forms.

It is necessary, however, not to overrate the effect of farmyard manure. The frequent addition of heavy dressings of stable manure for a generation or more can alter completely the character of a soil and this has happened in many market garden areas. It is quite common, for example, to find garden soils containing 15 or even 20 per cent. of organic matter whilst ordinary farm soils in the same area contain only 7 or 8 per cent. It is not to be expected that the limited amounts of manure available in ordinary farm practice can do much more than maintain the content of organic matter in the soil. The amount of organic matter in 10 tons of manure is about 2 tons which would increase the organic matter of the surface 8 or 9 inches of soil by only 0·2 per cent. The increase in humus would be very much less. Such differences are not measurable for they are not so large as the error incurred in careful soil sampling. The change becomes apparent only after repeated dressings.

The application of farmyard manure to grassland, not only has a fertilizing and water-retaining effect, but is also valuable from the way it acts as a mulch and affords the springing grass in the early months of the year some protection from cold and drying winds. At Rothamsted on the permanent grass plots it is often noticed that the plots which receive applications of farmyard manure once in every four years start a little earlier and make a quicker growth than the others. This mulching effect partly accounts for the great value attached to dung as a dressing for permanent grassland on open chalky soils, as in Wiltshire. Such a practice is wasteful of the farmyard manure as a fertilizer, for the loss of nitrogen is considerable, but the waste is tolerated in view of the gain to the physical or mechanical condition of the land.

Crop Response

It is not possible to make a strict comparison between farmyard manure and inorganic fertilizers. Many experiments have, however, been carried out in different parts of the world and the general conclusions are that, for equal amounts of nutrients, the manure is always less effective in increasing crop yield in spite of its other beneficial effects on the soil. Where the grower is aiming at a large crop it is more economical to supplement a moderate dressing of farmyard manure with a suitable fertilizer compound than to use a larger amount of manure alone.

The above observations on the fertilizing properties of farmyard manure are borne out by the average increases in crop yields obtained with fertilizers, with and without the addition of dung. We have already seen that the response to nitrogen is not greatly influenced by the presence of an average dressing of dung, whereas the response to phosphate is reduced by about 50 per cent. and the response to potassium by about 60 to 70 per cent. For 1954 prices, the most profitable dressings of nitrogenous fertilizers were only slightly reduced in presence of dung (by 4 or 5 lb. N per acre) whilst the reduction for phosphatic fertilizer was 45 lb. P_2O_5, and that for potassic fertilizer 67 lb. K_2O. This, in effect, means that the phosphorus and potassium compounds in farmyard manure are of greater immediate value than the nitrogen compounds, a fact which is commonly ignored. The use of dung enables the plant to respond freely to additional nitrogen and reduces the need for phosphatic and potassic fertilizers.

The average crop responses to 10 tons of dung per acre vary for different regions but for the country as a whole they amount to:

	Potatoes.	Swedes.	Mangolds.
With Fertilizers .	1·4	2·6	2·7 tons.
Without Fertilizers .	2·8	6·4	7·7 tons.

As percentage increases in yield, these figures are similar for all three crops but obviously the greatest financial benefit will be obtained by giving the potato crop first preference. This is, in fact, common farm practice; for example, in the arable districts of the Lothians the potatoes may receive about 15 tons dung per acre which leaves sufficient for only about half of the root break.

In ordinary mixed farming undoubtedly the best way of utilizing farmyard manure is to apply it to the root crops, and especially to mangolds and potatoes. Any surplus dung after the requirements of the root crops have been satisfied, is probably best given to the young seeds in the early winter, to act both as a fertilizer and as a mulch. The seeds benefit greatly, and at the same time much of the added fertility is retained for the corn crop that follows; manuring the young seeds is certainly preferable to the very general custom of manuring the old ley before it is ploughed up for wheat or oats. A certain amount of the farmyard manure made on the farm should, however, always be reserved for the meadow land, expecially on light soils and on land comparatively newly laid down to grass. Of course dung would be wasted on rich grazing land; it is the thin light soils that are cut for hay, or grassland that has only been laid down for a few years and has had no time to accumulate a stock of humus, which are most benefited by an occasional dressing of farmyard manure—once in every four or five years.

Farmyard manure has frequently been blamed for carrying the seeds of disease and of weeds, which have passed through the animals making the dung in an unchanged condition, and thus contaminating the land for other crops. When bullocks have been fed with swedes affected with "finger-and-toe" and the uneaten fragments of the roots have been thrown among the litter, the spores of the disease have been found to live unharmed through the making and rotting of the manure, so that fresh land may thus become infected when the dung is carried on to it. Similarly, when hop bines are used as litter, the spores of the hop mildew are not destroyed. As regards weeds, farmyard manure is very commonly employed for root crops, in which case the usual cultivations will keep down any weeds whose seeds were in the dung, and when the dung is put on grassland the weed seeds stand little chance of establishing themselves.

Cost of Farmyard Manure

What price should be set upon a ton of farmyard manure is a question often asked, but no general answer is possible, as so much depends upon the other conditions prevailing upon the farm. As a rule, farmyard manure is part of the normal output of the farm; the farmer has only to make it and use it to the best advantage, he is not concerned with the question of whether it would be cheaper to replace it with an equivalent amount of some other fertilizer. There are, however, occasions when the problem does arise of whether it is cheaper to make farmyard manure, to buy it, or to attempt to replace it by fertilizers; for example, the men who are farming specially for potatoes or hops often fatten bullocks or pigs solely for the sake of the manure thus made, and are content to lose money on the live stock because of the value of the dung. Since farmyard manure made in this way is often a very expensive article, it is important to try and put some monetary value on it, so that the farmer may attain a clearer idea of the profit or loss attached to the keeping of live stock as manure makers. It is, of course, possible to treat farmyard manure like any other fertilizer and value it on the unit system (see Chap. XIII) the result of which would be approximately as follows:

Farmyard manure contains:

	£	s.	d.
0·6 per cent. Nitrogen at 13*s*.	= 0	7	10
0·3 per cent. Phosphoric Acid at 7*s*. . .	= 0	2	1
0·5 per cent. Potash at 6*s*.	= 0	3	0
Value per ton	= 0	12	11

Much weight cannot, however, be attached to such a valuation, because the unit values are those for concentrated fertilizers and do not apply to dung. Considering the slow availability of much of the nitrogen in dung its unit value should be much below 13s. On the other hand, the organic matter supplied in the farmyard manure is not valued; yet it is for the effect of this organic matter on the texture of the soil that farmyard manure is most generally required. The cost of handling farmyard manure is much greater than that for an equivalent amount of fertilizer, and should also be taken into consideration but cannot well be estimated, because it will vary on each farm.

While it is thus practically impossible to value farmyard manure on its composition, a proper system of book-keeping will show what it costs to make, in a manner that is independent of the profit and loss upon the live stock. In this way a farmer can form for himself a clear idea of the economics of dung-making as compared with the purchase of either town manure or fertilizers. The most valid principle on which a cost can be worked out, and one which does justice equally to the live stock and to the manure, is to charge the dung with the cost of the litter and with the manure value of all the foods consumed in the yards or boxes. These manure values are what the valuer would allow to an outgoing tenant for the

TABLE 44.—COST OF MAKING ONE TON OF FARMYARD MANURE.

Foods.	Manure Value per ton.	Royal Agricultural Society's Farm, Woburn.						Cambridge University.				Wye.		Ordinary Farm.			
		1899.		1900.		1901.		1		2				1		2	
		Quantity.	Manure Value.	Quantity.	Manure Value.	Quantity.	Manure Value.	Quantity.	Manure Value.	Quantity.	Manure Value.	Quantity.	Manure Value.	Quantity.	Manure Value.	Quantity.	Manure Value.
	s.	Cwt.	s.	Cwt.	s.	Cwt.	s.	Cwt.	s.	Cwt.	s.	Cwt.	s.	Cwt.	s.	Cwt.	s.
Peas	30·0	...	...	...	...	...	...	...	...	...	...	...	...	...	...	18	27
Mangolds	2·5	...	...	...	...	...	...	120	15·0	120	15·0	...	...	...	...	120	15
Swedes	2·5	22·0	2·8	22·5	2·8	21·5	2·7	...	...	...	...	8·5	1·1				
Decorticated cotton cake	56·0	2·5	7·0	2·5	7·0	1·5	4·2	...	...	6·0	16·8	...	...	...	...	...	...
Undecorticated cotton cake	34·0	...	...	...	...	...	...	...	...	...	...	...	...	165	280	75	128
Linseed cake	38·0	...	...	...	...	...	...	...	...	...	...	1·6	3·1	260	494	170	323
Oats	14·0	...	...	...	...	...	...	...	...	...	...	...	...	90	63	200	140
Barley	14·0	2·0	1·4	...	...	...	...	...	...	...	...	...	...	...	...	...	...
Maize meal	14·0	...	...	2·1	1·5	1·5	1·0	...	...	...	...	...	...	140	98	180	126
Bran	30·0	...	...	...	...	...	...	...	...	...	...	...	...	5	8	20	30
Straw	7·0	6·0	2·1	8·5	3·0	4·8	1·7	...	...	...	...	...	...	...	...	...	...
Hay	15·0	...	...	...	...	2·7	2·0	10·0	7·5	10·0	7·5	3·5	2·6	1420	1065	1860	1395
LITTER:																	
Wheat Straw	30·0	10·0	15·0	10·0	15·0	9·5	14·2	17·5	26·3	16·5	24·8	6·0	9·0	1040	1560	1500	2250
Peat Moss	37·3	...	...	...	...	...	...	...	...	...	...	4·0	7·5	...	...	120	224
Total cost		28·3s.		29·3s.		25·8s.		48·8s.		64·1s.		23·3s.		3568s.		4658s.	
Quantity dung made		46·7 cwt.		49·5 cwt.		43·3 cwt.		100 cwt.		110 cwt.		90 cwt.		9200 cwt.		14,320 c.	
Cost per ton		12·1s.		11·8s.		11·9s.		9·8s.		11·7s.		5·2s.		7·8s.		6·5s.	

fertilizing material which he brought on to the farm during the last year of his tenancy and which he leaves behind in the form of dung.

The manure value of any food (Chap. XIII) is based upon its composition and represents the value at current market rates of whatever part of the food has a fertilizing value and may be supposed to find its way into the manure. To arrive at the cost of the dung, the manure value of all the food consumed is added to the cost of the litter and the sum is divided by the amount of manure ascertained to have been made. This principle was applied to the data obtained from some of the feeding experiments already quoted (Tables 33–35), and also to two cases extracted from the accounts of an ordinary farm (Table 44). The first column gives the nature of the food and the second its manure value per ton; the remaining double columns give for each food the amount consumed in the experiment and its manure value. The manure values are rather lower to-day but the original figures are given for the sake of the calculation in Table 45. Adjustments to current prices can readily be made (p. 309). The cost of the litter is set out below, and, added to the manure values, gives the total cost of the manure made in each case, the amount of which is also shown. Working on these lines we see that the farmyard manure cost from 5s. to 12s. a ton to make on the farm, without taking into account any profit or loss on the live stock, because this latter question is so much dependent upon the turn of the market and the skill of the dealer. It is necessary to discriminate and to keep distinct the two operations—the making of dung and the fattening of the cattle—so that a conclusion can be reached as to the profitableness of each separately. Of course, in making out the charges against the cattle, the whole cost of the cake, etc., which they consume must not be put down, but only that part of it which is not debited to the dung as manure value.

To make this clearer, we can consider the balance-sheet for the feeding of two of the heifers already mentioned:

TABLE 45.—CAMBRIDGE, No. 2.

Dr.	£	s.	d.	*Cr.*	£	s.	d.
Purchase price of 2 Heifers	30	0	0	Manure value of Mangolds . . .	0	15	0
6 tons of Mangolds at 5s.	1	10	0	Manure value of Hay .	0	7	6
½ ton of Hay at 45s. .	1	2	6	,, Cake (Charged to Dung)	0	16	10
6 cwts. of Decorticated Cotton Cake at £8 .	2	8	0	Sale price of Heifers .	34	0	0
Attendance, 12 weeks at 6d. . . .	0	6	0				
Balance, being profit .	0	12	10				
Total . . .	35	19	4	Total . . .	35	19	4

Thus the feeding resulted in a small profit of 12s. 10d., and at the same time, as was shown in Table 44, 5½ tons of farmyard manure were made at a cost of 11s. 8d. per ton, or if the heifers are considered to have been fattened solely for the purposes of making dung and the two accounts are combined by crediting the 12s. 10d. profit to the dung, the latter has cost about 9s. 4d. per ton. By this method it is possible to estimate what it will cost the potato or hop grower who sets out to keep cattle solely for the purpose of making dung. It is for him to decide whether he can secure sufficient profit from the cattle themselves to make it worth while to buy farmyard manure at such a price. A big cake bill is indeed a source of great loss on many farms, unless the cattle themselves pay for their food; the increased richness of the dung due to the purchased food will not produce a commensurate increase in the crops.

CHAPTER IX

WASTE ORGANIC COMPOUNDS

Animal residues, guano, fish guano, meat-and-bone meal, greaves, hoof-and-horn meal, poultry manure—Plant residues, rape dust and other cake residues, brewing and distilling waste products, seaweed, peat moss litter—Town refuse, screened dust, sewage sludge.

A GREAT variety of animal and vegetable substances are used as manures, as well as dung and composted straw. They are usually by-products of the slaughter-house or of factories but they also include bird excreta, sewage and seaweed. Dried blood and shoddy are essentially nitrogenous compounds fairly easily decomposed in the soil and have been described in Chapter III. The bone products are, in the main, phosphatic compounds and have been dealt with in Chapter V. The others, which contain nitrogen and phosphorus and sometimes potassium, are to be regarded as mixed organic fertilizers and are conveniently grouped together according to their source.

Animal Residues

Guano.—The term "guano" (Spanish *huano* = dung) is properly restricted to a fertilizing material consisting almost wholly of the excreta of sea-birds, which has accumulated upon certain oceanic islands where rain rarely falls. The original guano came from islands off the coast of Peru between the 7th and 20th degrees of south latitude, and this "Peruvian Guano" once formed the bulk of our importations. Other deposits, formed under similar conditions of climate and situation, have also been opened up. All these deposits, being of similar origin, possess many features in common. The islands, small and uninhabited, are the resort for breeding purposes of enormous flocks of pelicans, albatrosses, and other oceanic birds, which resort to land only in their breeding season. On the favoured spots they nest very closely, and the young birds are fed great quantities of fish brought by the parent birds. In addition to the excreta, the deposit thus contains many carcasses of young birds and fragments of fish, feathers, seaweed, and also some sand and stones. When the birds leave the island the tropical sun and the intense dryness of the atmosphere rapidly desiccate the accumulated materials and prevent any change or loss by fermentation.

The excreta of the birds is highly nitrogenous, consisting very largely of uric acid, together with a fair amount of phosphoric acid derived from the fish, which is the exclusive diet of the birds. An old analysis of a white Peruvian deposit, consisting mainly of recently deposited excreta, showed as much as 18·3 per cent. of nitrogen and only 9·2 per cent. of phosphoric acid (P_2O_5).

TABLE 46.—ANALYSIS OF FRESHLY-DEPOSITED GUANO.

Water	10·9
Organic matter and Ammonium Salts	65·63
Containing Nitrogen	(18·32)
Phosphoric Acid	9·20
Lime	6·08
Alkaline Salts	6·43
Sand	1·76

Dry as is the climate, a certain amount of change still goes on; the uric acid is fermented to urea and to ammonium salts, some of which are volatilized, while the occasional rains dissolve out both the ammonium compounds and soluble phosphates and the alkali salts. As a result, the composition of guano deposits is extremely variable, both in the different strata of one deposit and still more in passing from island to island. The older a deposit is, and the greater the washing it has received, the more will it have lost nitrogen and the richer will it have grown in phosphate, until from the material described above, deposits are formed containing little or nothing beyond calcium phosphate.

The analyses (Table 48) illustrates how great is the range in quality that thus results.

Peruvian guano is derived from three groups of islands off the coast of Peru, of which the most important, Chinchas, is a little south of Callao. The fertilizing value of the deposit was known long prior to the Spanish occupation of the country, in many parts of which crops could not be obtained without the aid of guano. A. von Humboldt was the first European to call attention to the use of guano; he brought samples home with him about 1804, at which time he found some fifty vessels annually employed in carrying guano from the Chinchas Islands for use in Peru.

The exportation to Europe, however, did not begin until nearly forty years later, the first cargoes being landed in Liverpool early in 1840. The success of the manure was rapid and the exportation soon assumed considerable dimensions, as much as 283,300 tons reaching the United Kingdom in 1845. One report refers to 300 ships crowding round an island, waiting for cargoes, in October, 1844.

During the earlier years Peruvian guano was derived from the Chinchas Islands only, and was an exceedingly rich deposit, containing 11 to 15 per cent. of nitrogen. Another of the islands, Ballestas, has yielded a very rich guano with more than 12 per cent of nitrogen and about an equal amount of phosphoric acid. It is difficult to form an adequate idea of the enormous bird population of these islands and the amount of food consumed during the breeding season, but a recent commission which visited the islands estimated the current production of fresh guano as 10,000 tons per annum.

Freshly deposited guano is light grey in colour; it becomes brown as it ages and undergoes some decomposition. A law has been passed by the Peruvian Government forbidding the working of the deposits during the breeding season. The guano islands are now, in fact, being regularly "farmed," and the exportations consist of the previous year's rich deposit, together with a certain amount of the older accumulated stock.

It is usually a loose dry powder, but there are found in it occasional fragments of slaty rock, with a number of half-decayed feathers in the richer specimens. It possesses a strong and characteristically ammoniacal smell and an alkaline reaction due to the presence of ammonium carbonate.

The following detailed analysis, Table 47, shows the

TABLE 47.—ANALYSIS OF CHINCHAS GUANO, 1897

Nitrogen as Nitrate	0·32	
,, as Ammonium Salts	3·94	
,, as Uric Acid	8·85	
,, in other Organic Forms	2·98	
Total Nitrogen		16·09
Phosphoric Acid soluble in Water	2·63	
,, soluble in Ammonium Citrate	6·29	
,, insoluble ,,	·37	
Total Phosphoric Acid, all soluble in 1 per cent. Citric Acid solution		9·29
equivalent to Tri-calcium Phosphate		20·28

composition of a sample of the Chinchas deposit; the nitrogen is present mainly in compounds soluble in water—uric acid, a little urea, guanine, and ammonium salts, with a trace of nitrate. The phosphate is completely soluble in dilute citric acid. It may be taken as a general rule that the lower the percentage of nitrogen, the less of it will be found in a soluble form, and the more insoluble will the phosphatic compounds be, so that the richest guanos are also the most readily available for the plant (Table 18). It is also characteristic of a good guano that the compounds of nitrogen present are very varied and undergo different changes in the soil before they become available, so that the crop is fed steadily and continuously. It is this property and the fact that guano is rich in phosphates as well as nitrogen, and also contains a little potassium, which makes it so popular. It is applicable to all crops but is always dearer than the inorganic fertilizers when valued on a unit basis.

Other deposits of guano which can be classed as nitrogenous are those from the islands off the south-west coast of Africa. There is a serious shortage of phosphates in South Africa, however, so that it is unlikely that there will be much available for export.

Fish Guano.—This is manufactured in many places where any considerable fish waste is available. The oil is extracted by heat and pressure, and the remaining material is dried and disintegrated as finely as possible. The nitrogen varies between 6 and 9 per cent. and the phosphoric acid represents from 13 to 20 per cent. of tri-calcium phosphate; availability depends upon the fineness of grinding. Fish guanos generally contain some oil which has not been removed in the process of manufacture, but this does not seem to delay the decomposition in the soil.

Fish guano is a comparatively active nitrogenous manure, since some of the compounds it contains are soluble in water and are rapidly decomposed by bacteria; the main constituents are, however, proteins which resist attack to a greater or less degree. In consequence, fish guano shares with the true guanos the property of continuing to yield nitrogen to the plant throughout the growing season, though the compounds in fish guano must be regarded as a little less active than those in Peruvian guano. Fish guano has for many years been a favourite manure among hop growers; it is also occasionally used for root crops when farmyard manure is not available. It should be applied early in the year, when the land is first worked, and should be dug or ploughed into the land as soon as sown, otherwise rooks and other birds will eat it as long as they are allowed to do so. Like all manures of this class, it is injurious to germinating seeds or the tender rootlets of growing plants, until it has been in the soil for a short time and the first active fermentation is over.

Meat Guano.—Meat-and-Bone Meal is prepared from all kinds of slaughter-house refuse in much the same way as fish guano—the waste of carcasses, condemned imported meat, tallow boilers refuse, the residues obtained in making meat extracts, and so forth, are heated and pressed to remove fat, and the residue is then finely ground. Material of this class, though more often after treatment with acid or other admixture, is known in America as "tankage." In some cases a good deal of bone is mixed with the material before grinding, and the resulting "guano" approximates to bone meal; in other cases the nitrogenous material predominates. Thus the nitrogen may be as high as 12–13 per cent. in which case there is little or no calcium phosphate, or it may be 4 to 5 per cent. with 35 to 40 per cent. of calcium phosphate.

TABLE 48.—COMPOSITION OF GUANOS AND KINDRED FERTILIZERS.

	Nitrogen.	Phosphoric Acid.	Equivalent to Tri-calcium Phosphate.	Potash.	Oil.
Peruvian Guano, Ballestas, 1902	12·24	11·36	24·76	...	...
" Lobos, 1902	1·50	31·63	68·95	...	...
" " 1906	8·4	13·17	28·70	2·85	...
" " 1906	2·37	21·61	47·12	2·90	...
Ichaboe . . .	8·64	12·14	28·09	2·47	...
Fish Meal	8·97	8·87	19·34	...	4·96
"	6·26	5·96	12·99	...	...
Meat Meal	12·3	0·92	2·0	...	...
"	6·36	14·32	31·27	...	...
Greaves	6·22	5·48	11·96	...	...
"	2·61	2·56	5·50	...	...
Dried Blood	9·65	0·83	1·82	...	...
Rape Dust	5·08	1·58	3·44	...	...
Damaged Cotton Cake . .	4·78	1·73	3·77	2·77	...

In its action and uses meat-and-bone meal is very similar to fish guano, and all that has been said about the time and manner of application of the one, equally applies to the other. Some growers appear to prefer fish to meat guano, but this is probably due to its more uniform composition. There is no evidence of the relative superiority of one over the other. The prices of the better grades are high because they are used as a cattle, and especially as a poultry food.

Greaves may be regarded as a low grade of meat-and-bone meal. It is the waste from tallow-making, and consists of the scraps of cartilage and bone which remain after the fat has been melted down and expressed as far as possible. The resulting waste material is still very fatty, and contains anything from 1·5 up to 9 per cent. of nitrogen, with phosphates varying from 5 to 12 per cent. of calcium phosphate. As a rule, the mechanical condition of greaves is bad and much against its proper distribution in the soil; the price is also often higher than its nitrogen content would warrant because reasonably clean samples can be used as poultry food. The amount of fat present is again possibly detrimental to its availability. Since greaves is extremely variable in its composition, according to the kind of material which happens to be treated at the factory from day to day, it is difficult to buy any large bulk on a guarantee, just as is the case with shoddy. It is difficult also to judge a consignment from a small sample, so that, as with shoddy, it is best to fix the price on the agreed unit value for nitrogen, taking the mean of several analyses from the bulk.

Hoof and Horn Meal.—This is still another animal waste product of the slaughter-house, for which there is a steady demand. It generally contains from 7 to 15 per cent. nitrogen which is slow acting because it is present mainly as stable epidermal protein. Hence it is not to be expected that enough nitrogen will be liberated for the plant in the season of application unless the meal is in a state of fine division and is applied in relatively large amounts to a soil in good condition. Sometimes the material contains a little bone so that there may also be present about 20 per cent. calcium phosphate. The output is small and the price is usually high per unit of plant food and obviously the material is of little use unless it is finely ground.

Poultry Manure.—In some areas of the country there are available appreciable quantities of poultry manure mixed with variable amounts of litter. The average recoveries in the excreta from laying birds of the nitrogen, phosphorus and potassium in the food are respectively about 70, 75 and 80 per cent. The recovery of nitrogen is greater than that from larger animals and of the total nitrogen in the fresh excreta, 60 per cent. is in uric acid and 10 per cent. in ammonium salts. The average percentage composition of the fresh excreta found during feeding experiments conducted for two years at Edinburgh is given alongside some American figures.

TABLE 49.—PERCENTAGE COMPOSITION OF POULTRY MANURE.

	Edinburgh.	New Jersey.	Pennsylvania.
Water	70	78	76
N	1·42	1·05	1·48
P_2O_5	1·16	0·82	0·96
K_2O	0·58	0·51	0·47

On a dry matter basis, all three contain about 4 per cent. P_2O_5 and 2 per cent. K_2O. The figure for nitrogen is 4·8 for Edinburgh and New Jersey, but 6·2 for Pennsylvania. Breed and age of bird and feeding, however, affect the results.

It will be observed that fresh poultry manure contains twice as much nitrogen and three or four times as much phosphate as farmyard manure; the figures for potassium are about the same. But, of course, allowance must be made for the inclusion of any litter—very often granulated peat which is a very inert material—and for losses during storage. The loss of nitrogen from a heap of wet manure may be as much as 60 per cent.; this can be halved by mixing the manure with litter and allowing it to dry in air. Admixture of superphosphate has been recommended to reduce loss of ammonia and calcium hydroxide has been reported as being a good deodorizer as well as a conserver of nitrogen. The fresh manure may also be artificially dried at 100°C without serious loss of nitrogen. The quantities available for sale are, of course, relatively small; as a source of nitrogen it is equal to other readily available nitrogenous compounds and it should be balanced by the addition of phosphate or potassium for certain crops and soils.

Plant Residues

Rape Dust and other Cake Residues.—In the manufacture of oil cake the oil-bearing seeds are subjected to great hydraulic pressure, either in bags or in metallic moulds which permit of the escape of the oil. The pressure is increased, aided sometimes by a little heat, until as much oil as possible has been obtained, there being left behind a cake consisting of the other parts of the seed, the proteins, carbohydrates, fibre, minerals, together with some residual oil. The cake is usually a valuable cattle food and is sold as such. In crushing rape seed, however, the resulting cake may be impure; rape seed not only contains a large proportion of impurities, but often also a good deal of wild mustard seed, from which, when the cake is used as food, mustard oil is generated in the stomach to a dangerous extent. Hence only the purer grades of rape cake are used for cattle feeding; in the other cases the cake is ground to powder and sold for manure. More recently a method of extracting the ground rape seed with oil solvents (the oil can be recovered by distilling off the solvent) has been generally adopted; the bulk of the oil in the seed is obtained in this way. The residue, which is really improved by the complete removal of the oil, is used only for manure.

Rape dust, as the ground rape cake is termed, has long been valued as a manure; William Ellis in 1735 speaks of oil cake with approval as one of the Hertfordshire "hand dressings" for corn, and at the time of the beginning of scientific agriculture in the second quarter of the last century we find that the use of rape dust had become fairly general throughout the eastern counties.

Rape dust contains about 5 per cent. of nitrogen, with such small quantities of phosphorus and potassium that it must in the main be treated as a nitrogenous manure. In its action it may be classed with the fish and meat guanos previously described, in that decomposition and nitrification is set up rapidly and continues throughout the whole season. It has been largely used in the Rothamsted experiments, and the results with barley and mangolds (Chap. IV) show that, nitrogen for nitrogen, it is almost as effective as sodium nitrate or ammonium sulphate. In these cases, however, the manure is applied year after year to the same land, so that the residues unused in the year of application accumulate for the benefit of the crop in future years; other experiments show that it is active enough to produce nearly its full effect in the first season. The organic matter of rape dust has a beneficial effect upon the tilth of the soil; on the Rothamsted mangold field, as has been pointed out earlier (Chap. VIII), the best results as regards plant establishment are obtained on the plot manured with rape cake. In general farming, rape cake has been found suitable for barley, and is highly esteemed by hop growers. Like all manures of its class, it should not in its fresh condition be put in contact with germinating seeds or young plants, probably because of the fungi and moulds with which it becomes permeated in the soil.

Other cake residues of a similar character come on the market from time to time in the shape of damaged cargoes of cotton, linseed, or other cakes, that have been spoilt for food by getting damp and heating or by the access of sea water. They should be judged on the same basis as rape cake and their value estimated from their analysis.

Brewing and Distilling Waste Products.—Malt culms, which are the rootlets of germinated barley and are broken off when the malt is dried, and spent hops each contain 3 or 4 per cent. nitrogen and 1 to 2 per cent. P_2O_5; the former also contain about 2 per cent. K_2O. They are useful local sources of organic material but cannot be expected to have much direct fertilizing value.

Seaweed.—Around parts of the coast, especially in the south-west and Channel islands, seaweed is an important manure and is collected after heavy weather and piled in heaps to dry and rot. It is also used in the fresh condition for the early potato crop in some parts. Many species are present in the material collected, but the most common are species of Laminaria, called drift-weed, tangle or kelp growing just below the low-water mark, and of Fucus or wrack found between low and high tide marks. The composition varies according to season, but is characterized by a relatively high content of potassium. The following figures give some idea of the average percentage of plant nutrients in different materials:

TABLE 50.—PERCENTAGE COMPOSITION OF SEAWEED.

Fresh Seaweed.	Water.	Organic Matter.	N.	K_2O.	K_2O in ash.
Laminaria { stems	82	12	0·2	1·8	29
Laminaria { fronds	75	20	0·3	1·2	23
Fucus . .	71	23	0·4	0·9	15

The stems of Laminaria are richer in potassium than the fronds and both are much richer than the different species of Fucus. They contain less nitrogen than farmyard manure and it is less available, but seaweed decays rapidly in the soil to release easily soluble nitrogenous compounds. Seaweed contains about twice as much potassium as farmyard manure but less than half as much phosphorus. It is free from disease organisms and weed seeds, and its lack of fibre makes it particularly suitable for sandy soils. Since it decomposes almost entirely to gases and soluble compounds it should not be exposed to prolonged weathering. It may be mixed with some absorbent material if it is not possible to apply it at once to the land. Fresh seaweed contains sodium chloride which is valuable for mangolds and sugar beet, but may be toxic to other plants. For most crops and soils it should be supplemented with a dressing of a phosphatic fertilizer.

Peat Moss Litter.—This material is being increasingly used in market gardens to build up the organic matter content of the soil. The light-brown fibrous peat is better than the dark-brown material. The latter has undergone a different kind of decomposition and is relatively rich in lignin; it is highly colloidal and very difficult to wet once it has dried and is more suitable as fuel. The coarse light-coloured fibrous material is still cellular, absorbs water more readily, and is a better medium for biological activity. It contains negligible amounts of phosphorus and potassium, and its nitrogen becomes available very slowly; moreover it is very acid and some form of liming material should be added in suitable quantity.

Town Refuse

Waste materials of various types have to be collected and disposed of in all towns and the possibility of their value for manurial purposes has frequently been considered. In the main they comprise the contents of dust-bins and street sweepings, night soil and sewage; slaughter-house refuse has already been dealt with.

Screened Dust.—In most towns the cleansing departments collect and remove the refuse from buildings and streets to a depot where paper, rags, bottles, metals and cinders are removed by hand, by magnet and by screening, and the sale of these materials is an important item in local revenue. The screened dust which passes a ½-inch riddle contains considerable proportions of small particles of unburned coal or cinders, and vegetable matter, the former predominating in the winter months, the latter in the summer-time. Such material is commonly used for levelling waste ground and so on, but it has also long been used on agricultural land in the neighbourhood of many towns. There is every reason to expect that it would improve the texture of heavy soils, and considerable quantities have been applied—usually at a rate of about 10 to 15 tons per acre—to clay soils of the Home Counties, to the Carse soils near Dundee and Perth, and to various areas in the Midlands. About 14,000 tons were used in 1943 in the Dundee area, where it is regarded as particularly suitable for turnips, beet and cabbages.

From considerations of the nature of screened dust it is not likely to give spectacular results as a fertilizer. Its composition is fairly constant for different towns when the results are expressed on a dry matter basis. The dried material contains about 30 to 40 per cent. organic matter, 0·7 per cent N, 0·3 per cent. P_2O_5, 0·2 per cent. K_2O, and 3 per cent. CaO. This is not greatly different from farmyard manure containing 75 per cent. water, but of course the comparison is not strictly valid because much of the organic matter in the dust consists of carbon compounds like coal which decompose extremely slowly in the soil. It may, of course, have a long-term effect in liberating plant nutrients and the fact that it contains small quantities of such essential elements as boron, cobalt, nickel, manganese and copper is worth bearing in mind. Numerous experiments in different parts of the country have demonstrated that it has little effect as a fertilizer in the year of its application, comparing badly with farmyard manure. Observations have pointed to a certain improvement in the braird of beetroot and mangolds, due to the addition of screened dust, without affecting the final yield to any extent, and that the calcium, phosphorus and potassium present may be of some use even although the nitrogen is not of much account. But experimental evidence on the possible residual value of the dust is not available.

In some cases the screened dust is mixed with richer manurial waste material. For example, it is sometimes used to absorb slaughter-house refuse, or it is mixed with garbage. At some places it is pulverized and treated with sewage sludge and the mixture allowed to ferment and mature for some months. Yet again stable manure or night soil may be incorporated. These mixtures will obviously contain more nitrogen and other nutrients, and are likely to be of much greater manurial value.

Sewage Sludge.—Since the process of digestion in man does not differ essentially from that of animals, the greater part, and in the case of adults the whole, of the nitrogen, phosphorus and potassium contained in human food is excreted in the urine and fæces. We have already seen that when plants are grown to feed animals, the nutrient constituents drawn from the soil are for the most part returned to the land; the only fertilizing constituents which leave the farm permanently are in the corn, wool, eggs, milk, and the fat stock for the use of man. Even of these the husk of the grain, the wool, the bones and hair find their way back to the land eventually, but under modern conditions the permanently valuable constituents of human food which pass into the excreta are then wasted agriculturally by being washed away into the rivers and sea. The waste is enormous; for example, it has been estimated as equivalent to considerably more than 100,000 tons calcium phosphate per annum in this country, and the losses of nitrogen and potassium are of the same order. The difficulty of preventing the loss lies in the fact that most of the methods for rendering serviceable the wasted material cost more than an equal amount of fertilizer from some other extraneous source.

Wolff and Lehmann have estimated (Table 51) the average composition of human excreta, and the average yearly output of each individual, from which it will be seen that neither

TABLE 51.—COMPOSITION OF HUMAN EXCRETA.

	Fæces.		Urine.	
	Per cent.	Lb. per annum.	Per cent.	Lb. Per annum.
Water	77·2	...	96·3	...
Organic matter	19·8	20	2·4	22
Ash	3·0	3·0	1·3	11
Nitrogen	1·0	1·04	0·6	6·9
Phosphoric Acid	1·1	1·3	0·17	3·2
Potash	0·25	0·3	0·2	3·4

urine nor fæces are particularly rich fertilizers. These are mean figures for all ages; for adults the annual quantities should be at least half as large again. But even taking high figures of about 12 lb. of nitrogen, 7 of phosphoric acid, and 5 of potash, they are worth respectively about 7s., 2s. 3d., and 1s. 4d., or 10s. 7d. a year in all when converted into a marketable fertilizer. Though for a large population the total waste may thus seem to be large, it is yet but a small amount to be set against the expense of dealing with such a quantity of low-grade material so difficult to handle.

Many attempts have naturally been made to utilize the fertilizing material contained in human excreta; in China it is applied fresh to the soil, being fetched daily by hand from the cities for that purpose. In the towns of Flanders and the north of France it was the custom to collect the excreta in large tanks, and after fermentation, to cart them out in a liquid form to the fields, though modern views on public health are rapidly getting rid of such practices. Almost the only method of getting human excreta back to the land cheaply and inoffensively is in houses or small communities where the "earth closet" system prevails. There the excreta are mixed with dry sifted earth, which deodorizes them quickly and completely, the mixture is removed daily to a heap under cover, and in a very short time it is so completely broken down by bacterial decay that it can be spread upon the land. In some inland towns a fairly concentrated manure is produced by evaporating off the water, sometimes with the addition of a little acid to fix the ammonia arising from the urea, sometimes with powdered turf, etc., to give the finished material a better mechanical texture. The following analysis shows the composition of one such manure:

Water	13·9
Organic matter	63·7
Containing Nitrogen	6·74
Phosphoric Acid	3·12
Potash	2·16
Insoluble Ash	3·45

The almost universal prevalence of a water-borne system of dealing with excreta puts an end to all such methods and intensifies the difficulty of saving the fertilizing constituents; town sewage now contains only about 40 parts of dry matter per 100,000. Where the conditions are favourable and the community has at hand a sufficient area of light, permeable land which can be cheaply graded and adapted to irrigation, then the sewage waters, either with or without a preliminary treatment to get rid of suspended matter, can be profitably utilized in raising crops. But light land, permitting of free percolation, is necessary, and it must not be overloaded with sewage but allowed intervals for aeration and oxidation, or else the surface becomes sealed with a layer of organic matter difficult to break down and both percolation and purification cease. An acre of land is capable of dealing with the sewage of about one hundred people.

The more common method of disposal now consists in separating the bulk of the solid material from the liquid by sedimentation. Sometimes this is accomplished by using precipitating agents like lime or iron and aluminium salts under such conditions as will produce a bulky precipitate which carries down the organic matter of the sewage. The sludge is then removed, pressed to remove excess moisture and sometimes dried for sale as manure. The presence of iron and aluminium compounds is a grave disadvantage, however, for they can render the soil phosphates inert, and soluble aluminium compounds are toxic to plants. It is much more common now simply to allow the sludge to settle in primary sedimentation tanks and submit the supernatant liquor to a further process of filtration or activation to obtain a secondary sludge. Activated sludge is obtained by aerating the liquor for a few days in presence of nitrifying organisms and then filtering off the sludge on ash beds. The clear effluent is fit to pass to rivers; it of course contains most of the soluble salts so that very little of the potassium in the original sewage is recovered and less than half of the nitrogen and phosphorus. Sometimes the primary sludge is allowed to dry in lagoons to a moisture content of 70 to 75 per cent.; the drying is quicker if the sludge is first allowed to undergo a "digestion" in presence of anaerobic methane-producing bacteria. A raw sludge in which little decomposition has taken place is difficult to deal with besides having an offensive smell. It is also gelatinous and difficult to dry. But if it can be stacked in heaps and aerated, some decomposition takes place and drying is more rapid on account of the rise in temperature. The amount of moisture may be reduced to about 20 per cent. at which point the sludge can be broken down to a state suitable for transport and application to the land.

The composition of sludge varies but the following figures give some idea of the percentage of the chief constituents in the dry matter:

TABLE 52.—COMPOSITION OF SEWAGE SLUDGE (Dry Matter).

Type of Sludge.	Organic Matter.	N.	P_2O_5.	K_2O.	CaO.
Raw lagoon	55	2·5	1·3	0·2	2·6
Digested—dried	40	2·4	2·6	...	...
Activated	...	6	4	...	...
Lime precipitated	32	1·1	1·7	...	38
Iron-aluminium	39	1·7	1·3	...	9

Raw sludges contain rather less organic matter than farmyard manure, about the same proportions of nitrogen and phosphorus but very little potassium. Digested sludges contain less organic matter, on account of the additional decomposition, and hence more inert mineral matter. Activated sludge is much richer in nitrogen and phosphorus. The chemically-precipitated sludges are naturally high in lime and the other precipitants used. There is more nitrogen in the organic matter of sludges than in that of farmyard manure and it is usually more available; in activated sludge, it is almost as quickly available as inorganic nitrogen. It is generally recognized that in this type of manure only the nitrogen in excess of about 3 per cent. in the organic matter is available to plants. Since the percentage of nitrogen in the organic matter of digested sludge is about 6, 3 per cent. may be taken as being available. This should be kept in mind in using sludge as manure, for a large dressing might easily supply an excessive quantity of nitrogen. It cannot be expected to exert such a beneficial action on soil texture as farmyard manure, at least on heavy soils, and its residual value is less. Nevertheless, it is a useful organic manure for farms within easy reach of a sewage works.

CHAPTER X

MATERIALS OF LESS DIRECT FERTILIZING VALUE

Lime, liming materials, oxide, carbonate, waste materials, soil acidity, effects of acidity, correction of acidity, chemical effects, physical action—Gypsum—Salt, cereals, roots—Silicates—Green manuring—Growth regulating substances.

LIME, gypsum and salt are three substances which are seldom described as fertilizers or manures but which are used in considerable quantities in agriculture with valuable results. Lime fulfils many functions. It reduces soil acidity or prevents it from developing and at the same time supplies the element calcium which is essential for plant growth; it makes clay soils more friable so that drainage and aeration are improved; these effects promote the activity of the micro-organisms which decompose the organic matter and liberate nitrogen and other plant foods; it guards against the formation of inert phosphates and reduces the damage done by the "finger-and-toe" organism. There is good reason therefore for describing lime—and this also applies to other "liming materials"—as a soil improver. Gypsum is not used much in this country but is of great value in parts of the world where alkaline soils develop. Its effect is mainly due to a chemical action which can bring about a beneficial change in the physical condition of these soils. Salt has long been regarded as a material having an indirect fertilizing value. It is perhaps more correct to regard it as a compound containing sodium which is a valuable nutrient for certain plants, although it is still not certain that it is essential to plant growth. The fact that 5 cwt. salt can help produce an equal weight of sugar by the sugar-beet plant is sufficiently important to justify its inclusion amongst the fertilizers. The effects of silicates, green manuring and growth-regulating substances are also discussed in this chapter.

Lime

When the value of lime became known is impossible to ascertain, but we find that the use of both lime and marl was recognized among the Romans. For example, Pliny writes: "There is another way of nourishing earth by earth, which has been found out in Britain and Gaul. It is thought that there is a greater degree of fruitfulness in this kind than in any other. It is a certain richness of earth, like the kernels in animal bodies, that are increased by fatness. The principal of those, reckoned the fat kinds, is the white; of this there are many. One very acrid, that has already been mentioned. Another kind of the white is like a soft clay. It is found at a great depth; the pits very frequently dug a hundred feet down, narrow at the mouth; but the vein, as in metals, widening within. This is chiefly used in Britain. It remains eighty years; nor is there an instance of any man laying it twice on the same field. The Hedui and Pictones manure their fields with lime, which is likewise found very good for olives and vines. All marl ought to be laid upon ploughed land, that its virtue may be the easier sucked in by the soil. A little dung should be laid on with it, particularly with that kind that at first is too hard, and does not dissolve well enough to nourish plants. Besides, of whatever kind it is, it hurts the soil, by its being new, and does not render it fertile till after the first year."

The regular use of some form of lime or chalk was part of the accepted routine of farming as early as we possess any records of British agriculture, and among the manures it figures in all books of the sixteenth and seventeenth centuries. In fact, "the black and the white," dung and lime, were the only manures employed by the great mass of farmers until well into the nineteenth century.

Liming Materials.—Lime itself, or quicklime, is obtained by the "burning" of any form of calcium carbonate, which occurs as limestone (either pure in the Mountain Limestone of Derbyshire and North Yorkshire, or argillaceous in the Lias); as chalk, and in the shell sand on various parts of the coast. The so-called "burning" consists in driving off by heat the carbon dioxide contained in the calcium carbonate. The resulting lime, known as quicklime, stone lime, cob lime or lime shells, is calcium oxide and combines readily with water developing much heat and falling down into a fine powder termed "slaked lime"; this slaked lime will then combine with the carbon dioxide present in the atmosphere to form calcium carbonate again. When lime is applied to the soil it is rapidly converted into carbonate. These changes may be expressed by the following equations.

1. $CaCO_3$ + heat = CaO + CO_2
calcium carbonate — calcium oxide — carbon dioxide

2. $CaO + H_2O = Ca(OH)_2$
water — calcium hydroxide or slaked lime

3. $Ca(OH)_2 + CO_2 = CaCO_3 + H_2O$

A pure form of slaked or hydrated lime is sold but it is an unnecessarily refined product for agricultural purposes.

Oxide.—The quality of lime varies considerably, according to whether it has been made from pure limestone or from impure forms containing some admixture of clay and sand. In the former case the result is a white, "fat," lime which swells considerably on slaking and falls into a very fine powder; the other grey or thin limes do not slake so readily nor swell much, they also contain a smaller proportion of free lime and are less valuable for agricultural purposes. In some parts of the country the limestone is dolomitic and contains considerable proportions of magnesium carbonate. On burning, magnesium oxide or magnesia (MgO) is produced, a compound which has a greater effect on soil acidity than an equal weight of lime. Under the present Fertilisers and Feeding

Stuffs Act, however, a lime is valued solely upon its content of free calcium oxide soluble in a sugar solution. This frequently underestimates its agricultural value because there may be calcium and magnesium compounds present which, although insoluble in this solution, break down easily in the soil and exert a neutralizing effect. At one time there was a great prejudice against the use of magnesium limes but there is no real evidence that they are harmful when used judiciously and indeed may be more useful than a pure lime when the soil is short of magnesium.

Lime ashes, which are to be had cheaply in the neighbourhood of the kilns, consist of the waste accumulating in burning the lime and are therefore mixtures of lime in a powder with the ashes of the coal employed. The percentage of lime may vary from 20 to 60 according to circumstances, so that the value of each lot must be judged by its apparent cleanliness and freedom from clinker.

Carbonate.—Since lime becomes calcium carbonate in the soil, obviously the same results would ultimately be obtained by applying the latter material; the main advantage in the use of lime lies in the very fine state of division into which it falls on slaking and the consequent good admixture with the soil that is effected. A finely divided calcium carbonate is obtained in many parts of the country in the calcareous marls which occur in beds sufficiently near the surface to admit of working, as in the New Red Sandstone formations in Cheshire, Worcester, etc., or the shell marls which occur in Norfolk. A true marl is a clay containing a variable percentage of calcium carbonate; it is specially valuable on sandy or peaty soils, not only for its calcareous matter but also for the clay, which improves the texture of the soil.

In other parts of the country, where the superficial formations resting upon the Chalk are devoid of calcium carbonate, it was formerly the custom to sink bell pits into the chalk rock, haul it up in baskets, and spread it upon the surface. In Hertfordshire, for example, this chalking was part of the regular routine of farming from the earliest times of which we have records, and from the analyses of the Rothamsted soils it has been ascertained that by the repetition of the process a hundred tons or more per acre must have been applied before the beginning of the nineteenth century, there being now present in the soil from 2 to 5 per cent. of calcium carbonate, all of artificial origin. Chalk was also formerly carried for considerable distances on to the clay formations—the London Clay, the Gault Clay, and the Weald Clay—that are contiguous, but the increased cost of labour has put an end to this practice. In some coastal areas, large amounts of shell sand rich in calcium carbonate are available. It is in good condition for spreading and is freely used as a liming material.

None of the other British limestones are sufficiently soft to allow of their direct application to the land with any prospect of their reduction to a fine state by the action of the weather, but many lime works have grinding plants and produce both limestone and chalk in a finely ground condition. The speed of action of ground limestone in the soil will obviously be governed by the particle size. It is decomposed by acidic materials; in the soil it is attacked by carbonic acid and converted into soluble calcium bicarbonate. Even the finest particles produced by grinding are not so fine as those produced chemically in the slaking and carbonation process, but nevertheless they react quickly. Hence, provided the ground limestone is well mixed and does not contain particles larger than about ⅛-inch, there will be both fine particles capable of rapid action and larger ones to act as a reserve. The larger and harder the particles are, the longer will they take to dissolve. It is actually a disadvantage, apart from cost, to grind limestone completely to a state of flour for then it is very difficult to distribute evenly. Chalk is easier to grind than limestone but has often to be dried first which involves an additional cost in production. A statement on the percentage of calcium carbonate and its equivalent calcium oxide is necessary in the sale of ground limestone and chalk; the fineness of grinding is also required in the case of the former.

Waste Materials.—Calcium carbonate is the chief constituent of several industrial by-products. At paper works it is precipitated in treating soda ash with lime to obtain caustic soda. It is therefore in a very fine state of division but, even after standing in heaps, contains about 40 per cent. of water which makes it troublesome to handle. In some works it is partially dried by artificial heat and becomes quite floury when the moisture content has been reduced to about 20 per cent. It has also been burned to give back calcium oxide; this may contain some silicate which, however, is also effective in neutralizing soil acidity. Paper works waste carbonate usually contains a little caustic soda but the amount seldom reaches 1 per cent. and is practically harmless.

A somewhat similar material is obtained at beet sugar factories in the purification of the sugar. It usually contains 30 to 40 per cent. water, 5 or 6 per cent. organic matter, and 0·1 or 0·2 per cent. nitrogen; the latter are probably not of much account, and the physical condition of the freshly produced material makes it difficult to spread. There is, however, no doubt about its value as a liming material, which increases as the moisture content is reduced. Soap and bleach and alkali works also have waste carbonates to dispose of and again it is usually a question of getting them into a suitably dry state for transport and spreading.

At one time all these carbonates could be purchased at a nominal price but not many farmers took advantage of this and large bings were built up at the works. During the liming campaign carried out in this country in recent years, considerable quantities have been transported and spread on the land by contractors. Some blast furnace slags are also effective liming materials although they consist mainly of calcium silicates and therefore do not come into the schedules of the Fertilisers and Feeding Stuffs Act.

Other waste products containing lime must be used with care on account of the possible presence of materials toxic to plants and animals. Tannery waste lime, for example, contains sulphides which should be oxidized by weathering before application to a growing crop. The by-product in the manufacture of acetylene from calcium carbide is mainly a wet slaked lime and it too should be exposed out-of-doors until poisonous impurities have been decomposed.

Gas lime is a greenish-yellow, evil-smelling substance obtained during the purification of coal-gas by its passage over trays of freshly-slaked lime, which absorbs sulphuretted hydrogen and other sulphur compounds from the crude gas. Various sulphides and partially oxidized sulphur compounds of calcium are formed, and these are to some extent attacked by the carbon dioxide of the air with the liberation of the original gaseous sulphur compounds. The main reactions on exposure to the air are carbonation and oxidation, so that eventually the material becomes little more than a mixture of calcium carbonate and gypsum. It is in this oxidized form only that gas lime should be applied to the land, unless the ground is badly infested with some insect pest which the raw sulphur compounds may check or destroy, and even then on light soils the fertility of the land may be impaired for some time. The crude material from the gas-works should be laid up in heaps, mixed with a little earth for a year or more, spread on the stubbles in the early autumn and then ploughed in.

Soil Acidity.—The soil contains a proportion of very fine particles. Whether they are derived from the minerals by weathering processes, clay, or by the partial decomposition of organic matter, humus, they possess colloidal properties which means that they have the power of absorbing hydrogen and bases like calcium, magnesium, potassium and sodium. When the colloid particle comes into contact with an acid (and soil water carries in solution carbonic acid produced by the biological activity within the soil), the hydrogen of the acid can displace part of the absorbed bases. In temperate latitudes, calcium is usually the predominating base and the exchange process may be written

$$\text{Calcium colloid} + \text{carbonic acid} \rightleftarrows \text{hydrogen colloid} + \text{calcium bicarbonate}$$

This reaction is reversible and the extent to which it proceeds in either direction depends upon the relative concentrations of the base and the hydrogen, the cations in the solution. For example, if the calcium bicarbonate is continually removed by leaching, the reaction will continue to go from left to right; but if calcium carbonate is added to the soil it will form bicarbonate and the reaction will tend to go from right to left. In a humid climate, the movement of water is mainly downwards so that calcium in solution is constantly being carried to the drains apart from the relatively small quantity absorbed by the growing crop. The loss to the drains varies considerably according to soil type and rainfall but is generally accepted as being equivalent to from 1 to 5 cwt. CaO per acre per annum. As explained in Chapter IV, ammonium salts increase the normal loss of calcium. Acid fumes from certain industries are also responsible for increasing the loss.

As a result of this normal weathering process, the ratio of hydrogen to calcium in the colloid increases; the soil is said to become unsaturated with bases and begins to behave like an acid. It may become so acid as to affect the growth of the plant. Several methods are used to measure the extent to which the process has gone. The concentration of hydrogen in solution may be determined and expressed on the convenient pH scale on which 7 represents neutrality and lower figures an increasing concentration of hydrogen. A soil in equilibrium with excess calcium carbonate is slightly alkaline with a pH value slightly over 8. Soils which do not contain free calcium carbonate gradually change to give pH values on the acid side of neutrality; these values are useful guides as to the suitability of the soil for different crops. Oats, rye, and potatoes are very tolerant of acidity and may grow perfectly well at pH figures between 4 and 5. Sugar beet and barley on the other hand are sensitive to acid conditions. It is impossible to give the precise pH figures at which they will fail, for other factors concerned with soil minerals and climate are involved, but sugar beet will usually not grow successfully when the pH value is much below 6.

The usefulness of a pH determination must not be exaggerated, however, because it is only a guide to the condition of the soil. It is better to know either how much exchangeable or replaceable calcium remains in the soil or how much lime is required to replace the absorbed hydrogen which is responsible for a low pH value. A low pH value is much more serious on a sandy soil than on a heavy loam for the former contains much less colloidal material and therefore a smaller reserve of absorbed bases. On the other hand, less lime will be required to reduce the degree of unsaturation and raise its pH value to a suitable point. All plants require calcium—even potatoes may suffer from a deficiency on soils which have been allowed to become too acid—and the "lime-requirement" of an acid soil takes into consideration the crop rotation and farm practice of the area in question.

Effects of Acidity. The occurrence of acid conditions is usually revealed by the presence of plants which are tolerant of acidity and able to crowd out a sensitive crop which is weak and patchy. Sheep's sorrel (*Rumex acetosella*), spurry (*Spergula arvensis*), or corn marigold (*Chrysanthemum segetum*), become troublesome on acid arable soils. Leguminous crops like beans, peas and clovers commonly fail through lack of sufficient available calcium, and reference has already been made to the importance of lime for sugar beet and barley. Permanent grassland develops a "mat" of acid organic matter which acts like a sponge so that the roots do not penetrate into the mineral soil, and the vegetation becomes characterized by a predominance of fine-leaved species and an absence of clovers. Bracken (*Pteris aquilina*) and foxglove (*Digitalis purpurea*) are common on waste ground whilst moorland soils acquire a flora typical of the particular soil type which has developed. It must not be supposed that the use of fertilizers which contain calcium compounds will obviate the necessity of liming. Basic slag and calcium cyanamide are of assistance in reducing soil acidity somewhat, but the amounts used cannot be expected to take the place of the lime necessary to maintain a satisfactory base status of the soil.

Some micro-fungi are favoured by an acid reaction in the soil, and we find that various fungoid diseases of plants are specially prevalent on soils devoid of calcium carbonate, a notable example being the slime fungus, *Plasmodiophora brassicæ*, which causes "finger-and-toe," "club-root," or "anbury" in turnips, cabbages, and other brassicæ. This disease is uncommon on calcareous soils and can be reduced on soils where it does occur by a thorough liming, both after the removal of one turnip crop and again immediately before another is sown. As long an interval as possible should elapse between the two cruciferous crops to enable the spores of the disease to die off in the soil. The relationship between liming and the control of this disease is not thoroughly understood, however, for experimental results show great inconsistency. Degree of soil alkalinity (pH value), excess of lime, susceptibility of crop, physical condition of the soil and the possibility of different strains of the organism all seem to be involved.

On the other hand, alkaline conditions produced by liming favour the disease fungi responsible for scab on potatoes. Common scab is more frequent under dry conditions, corky scab under wet conditions. In both cases some control of the trouble is obtained by a substantial dressing of flowers of sulphur (500 to 1,000 lb. per acre) from which sulphuric acid is produced; the temporary acidity helps to secure a cleaner crop. Recent work suggests that an abnormal ratio of available potassium to calcium, which may be corrected by more acid conditions, is responsible for the incidence of those diseases. The turning in of vegetable refuse or green manuring is also found to be beneficial in the control of common scab.

Correction of Acidity.—The application of lime or some other liming material is an essential item in good husbandry except to soils which are naturally rich in calcium carbonate. At one time it was customary to give very large dressings at lengthy intervals. This is liable to upset the balance or availability of plant nutrients and it is better to apply about a ton per acre regularly at suitable points in the rotation—

preferably before the turnip or beet crop. All liming materials will eventually have the same effect if they are in a suitable condition for proper incorporation in the soil and are applied at rates which supply equal amounts of calcium. But it is a waste of material and effort to attempt to build up a reserve in the soil because the rate of loss to the drainage is greater the richer the soil is in calcium carbonate. It is sufficient to reduce the acidity to a point at which sufficient calcium will be available for the crop being grown. The effects of lime on the yield and quality of crops vary between wide limits so that individual results may be misleading. In some cases a moderate dressing may save a crop from being practically a complete failure, at other times the response will be less striking but nevertheless quite profitable. Its value to stock by raising the level of calcium in grass and encouraging clovers on grazing land is of outstanding importance. Many of the soils in Great Britain—over 40 per cent. in some regions—are deficient in lime and are failing to give the best return for the fertilizers they receive.

With respect to the choice of liming material, it is most often a question of deciding between lime, ground or unground, and ground limestone. Bearing in mind that one ton of pure lime or calcium oxide will have the same effect as about 1¾ ton (4,000 lb.) of calcium carbonate, it is easy to compare in the first instance the relative costs of purchase, transport and application. Ground lime is difficult to store on account of its readiness to absorb water, which causes the bags to burst, and is unpleasant to handle on account of its caustic character. It may be evenly spread by machine but until it is harrowed into the soil it is liable to cake and so lose the advantage of the fine state of division of the slaked lime. Lump lime has, of course, to be laid out in small heaps and allowed to slake before it can be spread. This method of liming is a very popular one but frequently leads to bad distribution even under the best conditions and, when wet weather prevails, the heaps become pasty so that even spreading is quite impossible. Ground limestone must be used in much larger quantities but it is easy to store, as easy to distribute evenly as ground lime, and has no burning action. Particles which are small enough to pass a 100 mesh sieve react almost as quickly as ground lime. If about half of the material is ground to this degree of fineness, the immediate effect on the soil will often be adequate for the next crop while the coarser particles up to about $\frac{1}{8}$ inch. in diameter will slowly exert their action in the following few years. The same observations apply to ground chalk and shell sand, but they are not so appropriate to waste carbonates unless the latter are dry enough to be evenly spread.

When the liming material is being used to correct serious acidity, rather than as a routine measure to maintain fertility, it should be applied as soon as possible after the ground has been cleared of the preceding crop and ploughed. It should then be worked into the surface soil so as to obtain a thorough incorporation. On grassland, lime is best applied at a time when a temporary check in growth will be least serious, but the carbonate is safe to apply at any time. It is not possible to prescribe suitable dressings without knowledge of the conditions in question; heavy dressings may be required to correct serious sourness but, as has already been indicated, it is better to apply moderate amounts of the order of 1 ton of lime or 2 tons of carbonate every 5 or 6 years than 5 or 6 times these quantities to last a generation.

Chemical Effects.—On the chemical side one great value of lime lies in its ability to keep the soil reaction suitable for the development of those bacteria which oxidize the organic compounds and release plant food. In the absence of lime, organic matter by its decay gives rise to various acid bodies which may be grouped as humic acid, and the action of many of the valuable groups of bacteria, such as the *Azotobacter* which fix nitrogen, and the nitrifying bacteria which convert ammonia into nitrates is inhibited. It has been shown that in soils that are acid through the accumulation of humic acid, nitrification is at a standstill and bacterial life generally is repressed in favour of the growth of moulds and micro-fungi, which compete actively with the crop for the plant food in the soil.

On all land which has been enriched by the residues of past manuring or by the debris of previous vegetation, lime is very necessary to promote the oxidation of the nitrogen compounds and the formation of nitrates for the crop; consequently it is on bog or peaty land, on old turf or reclaimed forest land, or on old gardens, that liming exercises its maximum effect.

But the nitrogenous compounds in the soil are not the only ones rendered more available by the presence of calcium carbonate; both phosphate and potassium are thereby kept in or brought into a more soluble form. When soluble phosphates are applied to the land they are quickly precipitated as di-calcium phosphate, ferric phosphate, or aluminium phosphate; on soils containing adequate amounts of calcium the first will predominate; the other two are formed where calcium is lacking. Now the effective solubility of the two latter phosphates in the soil water is much less than that of the precipitated calcium phosphate, and consequently the plant may remain short of phosphorus even though large amounts have been applied to the soil. Applications of lime or calcium carbonate are of great value on these soils partly because they may form a certain amount of calcium phosphate by interaction with the iron or aluminium phosphates, but mainly because they decrease the acidity of the soil and ensure the absence of soluble iron and aluminium salts which form very inert phosphates. The response to phosphatic fertilizers on a very acid soil is often small or negligible; if the acidity is not at least partially corrected the phosphate is fixed and lost to the crop. Furthermore, the soluble aluminium salts, sometimes present in very acid soils, are toxic to plants.

The action of lime upon the potassium compounds in the soil is also important. The soil water carries down the dissolved calcium bicarbonate which reacts with the colloidal particles. Hence some of their absorbed bases, potassium among them, are displaced by the calcium and so come into solution. Thus calcium is precipitated and potassium is found in the soil water. The action is the reverse of that which takes place when potassium salts are applied as fertilizers; whatever base is in excess in the soil solution will replace the others and be absorbed by the soil colloids. This base exchange reaction is analogous to that governing the reversible exchange of calcium and hydrogen dealt with in the discussion on the development and correction of soil acidity.

$$\text{potassium colloid} + \text{calcium salt} \rightleftarrows \text{calcium colloid} + \text{potassium salt}$$

Hence, lime may act indirectly by releasing for the plant some of the potassium absorbed by the soil.

That lime acts in this fashion may account for its beneficial effect upon clovers and other leguminous plants in a mixed herbage, or for the remarkable power of basic slag to promote the growth of white clover in a pasture where it was formerly dormant. Other phosphatic manures have often little effect in such cases, so that the lime action of basic slag, by liberating potassium, may be an important factor in the growth of the clover as well as the phosphate. If the basic slag is applied to a soil poor in potassium it has little effect. Again, after two or three applications to grassland it may cease to show its previous beneficial action upon the clover, because the readily replaceable potassium in the soil has been brought into solution and lost, so that a direct application of potassium salts becomes necessary.

The fact that a solution of calcium bicarbonate will react with potassium-containing silicates in the soil, so as to bring some of the potassium into solution, is only a particular case of a more general proposition applicable to any soluble salt. Whatever the salt may be, it will effect an exchange with the bases in the soil to a greater or less degree according to the relative amounts of salt and colloid. In practice the application of any soluble salts of calcium, magnesium, potassium and sodium to the soil is usually too small to bring about substantial changes in the composition of the soil solution, but that such changes take place in extreme cases is shown by the figures in Table 53. These show the average composition of the wheat straw ash for the 10 years 1882–91, the straw ash alone being given because there was practically no variation in the composition of the ash of the grain. The experiment was started in 1852.

TABLE 53.—PERCENTAGE COMPOSITION OF THE ASH OF WHEAT STRAW. MIXED SAMPLES REPRESENTING 10 YEARS, 1882–1891 (Rothamsted).

	Ammonium Salts and Superphosphate with:				
	0.	Sulphate Soda.	Sulphate Potash.	Sulphate Magnesia.	Sulphate Soda, Potash, and Magnesia.
Plot	11	12	13	14	7
Ash (crude) in Dry Matter, per cent.	5·84	5·69	5·93	5·52	5·89
Iron Peroxide, etc.	0·43	0·33	0·34	0·41	0·50
Lime	9·14	7·73	5·39	7·70	5·69
Magnesia	2·25	1·92	1·53	2·46	1·76
Potash	9·91	14·68	23·28	14·87	25·89
Soda	0·58	0·57	0·03	0·33	0·21
Phosphoric Acid	4·26	3·65	3·39	3·87	3·82
Sulphuric Acid	5·44	5·33	5·07	5·31	5·41
Chlorine	1·66	2·89	5·61	2·81	6·60
Carbonic Acid	trace	none	none	trace	...
Silica	65·19	61·93	54·26	61·06	49·68
Sand	1·46	1·43	1·76	1·39	1·32
Charcoal	0·06	0·19	0·60	0·42	0·61
Total	100·38	100·65	101·26	100·63	101·49
Deduct O=Cl.	0·38	0·65	1·26	0·63	1·49
Total	100·00	100·00	100·00	100·00	100·00

There is no great variation in the percentage of ash in the dry matter, but while the percentage of potash in the ash of the straw from Plot 13, receiving only potassium, is 23·28, it is raised to 25·89 when sodium and magnesium are also added to the potassium in the manure, but sinks to 9·91 when all the alkaline salts are lacking. From this low figure of 9·91 the addition of sodium causes a rise to 14·68, of magnesium to 14·87; neither the magnesium nor the soda in the ash are perceptibly raised. Hence, we may conclude that sodium or magnesium sulphate has an indirect effect upon the crop. These effects are of course dependent upon a series of related reactions in the soil and are not so apparent under normal farming conditions. As we shall see, sodium is a direct nutrient for some crops and can take the place of potassium to a certain extent without altering the uptake of potassium by the plant to an appreciable extent.

Physical Action.—As a result of the chemical reaction between calcium and the soil colloids, lime exerts a most important effect upon the texture of the soil. It acts by encouraging the finest clay particles to flocculate or form aggregates of larger units and so making the soil more permeable. It also preserves the calcium colloids which are much less sticky and water holding than sodium colloids. The soil thus becomes less retentive of moisture; its plough draught is reduced; percolation is increased, making the limed land dryer and warmer, so that it admits of cultivation earlier in the spring and is far more friable when dry; in dry seasons a clay soil will crack less and the crop will keep on growing longer, because of the improved texture of the soil.

It is difficult to exaggerate the improvement that lime effects in the dryness and workability of strong soils, which in many cases would not be fit for arable cultivation without adequate liming. It has already been mentioned that on the Rothamsted estate the custom of chalking has added from 2 to 5 per cent. of calcium carbonate to the surface soil, which is otherwise non-calcareous; but on one of the fields, formerly under experiment, the treatment had never been carried out. This field, Geescroft, formerly carried experimental crops of oats and beans, but during the rainy season about 1879 the land lay so persistently wet in the spring that on several occasions a tilth could not be obtained in time for sowing, and the land had to lie fallow, until at last cultivation was abandoned and the field was allowed to fall down into grass. Even then the herbage was very inferior and showed the wet character of the soil by the prevalance of *Aira cæspitosa*; yet in situation, drainage, and mechanical composition this soil was in no respects different from that of the other Rothamsted fields. The essential factor which caused all the difference in the character of the two soils was the absence of calcium carbonate from the Geescroft field, which for some reason had escaped the chalking given to the other fields. The physical improvement of a clay soil by lime is not apparent at once but grows from year to year after the application.

On the lighter soils—the sands and gravels—lime exerts a good effect by forming a weak cementing agent and increasing the cohesion of the particles. As a rule, however, it is not wise to apply quicklime in large quantities to light open soils, because oxidation of the organic matter may be pushed on too rapidly. It is safer to use some form of calcium carbonate.

The whole of the above discussion on the effects of lime points to the fact that its value in agriculture can hardly be over-emphasized. Liming used to be much more general than in recent years; there has been a revival of the practice but there is much leeway to make up, large deficiencies to be made good. The increase in the use of fertilizers makes the problem more urgent because bigger crops mean that more calcium is being removed from the soil and some fertilizers increase the loss of calcium in the drainage. Even better crops, in yield and quality, will be secured when the shortage of lime is rectified and the potential fertility of the soil fully realized.

Gypsum

That gypsum, or hydrated calcium sulphate, $CaSO_4, 2H_2O$, had a beneficial effect upon certain crops and soils has been known for a very long time; it was probably familiar to the Romans, and the knowledge survived to a certain degree among the southern nations, especially in connection with vines. In Britain it appears to have been less commonly used and no very general agreement as to its value had become traditional; in fact it was only among the hop growers of Kent and especially of Sussex, that any regular use was made of gypsum. In this latter case it is not easy to make out to what extent its employment was the result of experience, or of a quasi-scientific opinion which traced a connection between the action of sulphur upon the mildew of the hop, a supposed lack of sulphates in a mildewed leaf, and the sulphates in the gypsum.

In the latter part of the eighteenth century the value of gypsum for leguminous crops like clover and lucerne became widely recognized; Benjamin Franklin, in America, is said to have sown gypsum so as to form the word "plaster" on clover crops by the wayside, in order that passers-by should learn by the eye what had so stimulated the growth. It is, in fact, on leguminous and such other crops as are specially dependent upon calcium, that gypsum has an effect. Boussingault showed that when clover was manured with gypsum and improved thereby, the ash of the crop contained a greater proportion of potash but showed little change in either the lime or the sulphuric acid (Table 54).

TABLE 54.—PERCENTAGE COMPOSITION OF CLOVER ASH.

	Without Gypsum.	With Gypsum.
Potash	23·6	35·4
Soda	1·2	0·9
Magnesia	7·6	6·7
Lime	28·5	29·4
Oxides of Iron and Manganese	1·2	1·0
Chlorine	4·1	3·8
Phosphoric Acid	9·7	9·0
Sulphuric Acid	3·9	3·4
Silica	20·0	10·4

It has also been found, in testing the action of gypsum upon hops, that it has a beneficial effect only upon the soils where hops also respond to dressings of potassium salts, and that the result of applications of gypsum is similar to that of potassium salts, though to a less degree.

Gypsum is the only practicable material that can be employed to restore the workability of a heavy soil that has had its texture destroyed by long flooding with sea water. During the immersion, the salt in the sea water turns out the greater part of the calcium in the clay colloids and replaces it with sodium, and the resultant sodium clay is excessively sticky and retentive of water. Gypsum effects the reverse action and restores the calcium, and has been successfully used in the Netherlands and the south east of England to reclaim land flooded by the sea.

Gypsum has also been used in the reclamation of "black alkali" soils. These soils develop under semi-arid conditions in which sodium salts have produced a sodium clay and this has become impermeable to water through the removal of excess salts. Some sodium carbonate has been formed from this clay and dissolves the humus to give a highly-dispersed black layer on the surface of the soil. The presence of sodium carbonate also makes the soil quite alkaline. If free drainage can be obtained, it may be expected that the addition of gypsum will help to convert the sodium clay back to a calcium clay, but the impermeability of the soil is often such a serious handicap that the treatment is unsuccessful.

Salt

The uses of salt or sodium chloride in agriculture have been recorded from time to time for centuries and until the middle of last century its value as a manure was unquestioned. When cheap potassium salts became available, interest in the use of common salt practically ceased, for it was a constituent of most of the potassic fertilizers.

Cereals.—There is not much experimental evidence on the value of salt as a fertilizer for cereal crops. It has been a fairly common opinion that kainit, which contains a large proportion of salt, is particularly useful for barley grown on open thin soils or on the chalk soils of the eastern counties, and it has been shown that a chloride was somewhat better than a sulphate in the production of grain of good malting value. Generally speaking, the results show that barley may respond as well to sodium chloride as to potassium chloride, but that the former is not so effective for wheat.

Roots.—Much more information is available about the effect of salt on root crops. Its use in growing mangolds is a very old farming practice and gives substantial increases in yield. Care in application is necessary to avoid any delay in germination or damage to the young seedlings, and since salt, on account of its usual moist condition, is difficult to mix with other fertilizers and is therefore handled separately, the best plan is to apply it sometime before drilling the seed. The response to the common dressing of from 3 to 5 cwt. depends upon whether farmyard manure or potassic fertilizers are also used. It is greater in absence of these sources of potassium and amounts to 3 or 4 tons of roots per acre. The long-term mangold experiment at Rothamsted has shown that salt has been as effective as potassium sulphate over a period of nearly 60 years.

Salt has also been found to be beneficial to the sugar beet crop and often as effective as potassium chloride. In peace time this was not regarded as of much importance, because potassium fertilizers were reasonably cheap and could more readily be compounded with other fertilizer salts. But during the war, when supplies of potassium salts were difficult to obtain and a large proportion of what was available was required for the potato crop, the value of sodium as an alternative to potassium for the beet crop was rapidly realized. Many precise experiments have been carried out in recent years with the beet crop on a variety of soil types, and a mass of valuable data has been collected. The full results of this co-operative work have been of the greatest value for they all point to the fact that sodium is an actual plant food for the beet plant and may be as effective as potassium in increasing the yield of sugar; in fact, the response to potassium is quite small when salt has been applied. This result, as might be expected, is most consistent on soils which are relatively deficient in available potassium, for salt does not affect to any extent the absorption of potassium by the plant; it seems rather to fulfil the same function as potassium in the plant. The general recommendation now made is that salt should be used at the rate of from 3 to 5 cwt. per acre. Such a dressing will on the average increase the yield of sugar by from 3 to 5 cwt. per acre, provided the crop is adequately supplied with nitrogen. The ability of the beet plant to utilize salt in this manner is one of the most striking features of these investigations.

A criticism commonly levelled at the use of salt is that it spoils the tilth of heavy soils, making them more sticky and difficult to work. The explanation of such a result would be that the sodium chloride reacted with the clay particles in the soil to produce a sodium colloid, which, as previously mentioned, may become deflocculated so that the individual particles are dispersed. The sodium colloid is also associated with a lot of water so that the soil becomes impermeable and it is difficult to obtain a tilth. Although this may indeed happen with some soils—probably where a heavy dressing of salt has been allowed to react with the surface layer of soil—it is not a general occurrence and need not be feared if the salt is used in moderate amounts and incorporated with the cultivated soil.

Silicates

Silica is so large a constituent of the ash of many plants, particularly of the straw of cereals, that it was inevitably regarded as a necessary constituent of the food of such plants, and was naturally enough supposed to contribute to the stiffness of the straw. In his manures Liebig supplied the alkalies combined with silica, and when Way discovered that certain strata of the Upper Greensand, near Farnham, contained considerable quantities of silicates readily dissolved by acids, the rock was for a time extracted and ground up as a manure for cereals. But Sachs showed that these plants, however rich in silica their ash was when they had grown on ordinary soil, could yet be grown with complete success in a water culture devoid of any silica, and Jodin succeeded in raising four generations of maize in water cultures with no more silica than was contained in the original seed. It was also shown that the stiffness of the straw depended upon such physiological factors as light and exposure, rapidity of growth, etc., and was independent of the amount of silica present, so that the use of silicates for manurial purposes ceased. However, it must not be supposed that so large a constituent of a plant's ash is entirely without physiological function, and many experiments have shown that soluble silica does play some part, at present not properly understood, which enables the plant either to obtain more of the dormant phosphate in the soil or to utilize silica in place of phosphate. The silicates, however, possess no practical use as fertilizers, except possibly in some blast furnace slags whose "lime-value" lies in their easily decomposable calcium silicates.

Green Manuring

Green Manuring consists in the ploughing under of some rapidly growing crop—mustard or tares in this country, lupins on sandy soils on the Continent, and cowpeas in America being among the plants most commonly employed. The practice has three objects:

(1) The improvement of the texture of the land by increasing the store of humus; this is particularly valuable on heavy clays and on the light sandy soils at the other end of the texture scale.

(2) The saving of the store of nitrates, which on light warm soils form with great rapidity after harvest, and which may then easily be washed away. If some catch crop like mustard is sown immediately the stubbles are clear, it will grow with great rapidity after the first rain and will gather up these

nitrates, converting them into proteins, which become more slowly available on the decay of the plant material.

(3) For cleaning purposes; when the land is in very foul condition a good many weeds can be got rid of by growing a smothering crop.

On many soils green manuring may be extremely valuable, especially where there is any shortage of farmyard manure; a green crop of mustard turned in, especially if it has been previously manured with some fertilizer mixture, will have a lasting beneficial effect. Of course the "seeds" crop in the rotation has much the same effect, because of the roots and stubble left behind, but it does not always come round often enough in the rotation to keep the land in condition.

When vetches, lupins, or other leguminous crops are grown, the land is also enriched by the nitrogen gathered from the atmosphere by the bacteria living in the root nodules. Large areas of land in Pomerania and East Prussia have been brought under cultivation from the state of barren sandy heath, by ploughing in lupins manured with basic slag and potash salts, until a soil had been built up.

Curiously enough, on the sandy soil at Woburn, Voelcker always obtained better crops after mustard than after vetches, despite the fact that the vetches had contributed a greater weight both of dry matter and nitrogen to the land. The vetch compounds may decay the more slowly, but Voelcker further showed that the land was left drier by the vetch crop; that this was the cause of the superiority of the mustard as a green manure is rendered more probable by the fact that the result was reversed on the strong Rothamsted soil, where the vetches were the better preparation for a succeeding wheat crop. The real difficulty experienced in utilizing green manuring and catch crops generally on many soils in this country, to which they are otherwise most admirably suited, is the way they deplete the water supply for the succeeding crop. For example, a crop of vetches or crimson clover may be sown on the stubble in August or September and harvested in May, in plenty of time to prepare the land for turnips, but in many cases the soil and subsoil will be left so dry that the turnip crop will fail or be greatly reduced, unless the incidence of rain be unusually favourable. The difficulty of starting the catch crop after the drying effect of the harvested corn, and the dryness of the land which again ensues after the catch crop in spring, form the great objection to catch-cropping, which indeed only flourishes where the annual rainfall is well over 30 inches. In this respect mustard is the least objectionable crop, since it may grow in six or eight weeks under good conditions in autumn, and can then be turned in, leaving the ground broken to catch the late winter rainfall.

One method of cleaning market garden land and building up the organic matter of the soil has been developed with great success by Scarlett in the Inveresk area. The land is ploughed and tares are sown in early spring. At flowering in June the crop is crushed by a heavy implement and treated with 3 to 4 cwt. of calcium cyanamide per acre. The heavy decomposing mass is ploughed under after a week and the surface immediately prepared for a sowing of rye. This grows rapidly, for there is plenty of available nitrogen, and provides a second large amount of roots and foliage to be ploughed down in late autumn; the ground is ready for normal cultivation in the spring. Scarlett regards the complete destruction of perennial weeds and the reconditioning of the soil as adequate compensation for the loss of a crop which would probably be a poor or expensive one on account of the weeds; and he has estimated the cost of the process as being less than that of a good dressing of dung.

On light soils it is more general to fold sheep on the catch crops than to plough them in, and though a large part of the humus is thus lost to the land, there is still a considerable gain, while the essential manurial substances—nitrogen, phosphorus and potassium—are almost wholly returned to the soil. Where the land is light enough to be improved by the treading of sheep and the rainfall admits of catch-cropping, there is no better way of building up a fertile soil than by folding; the actual enrichment of the soil can be effected either by manuring for the catch crops with inorganic fertilizers or by feeding cake and corn to the stock. The losses inherent in making dung are thus obviated; the tilth of the land is improved by the humus and the trampling of the sheep; no more effective or cheaper system of growing corn can be devised than to alternate it with green crops consumed on the land, as is practised with so much success on the brick earths of West Sussex and the chalky loams of Wiltshire.

Growth Regulating Substances

Reference has already been made in the discussion on farmyard manure to the indole derivatives which it contains and which have the property of regulating plant growth. Many substances are able to promote growth in plants; they are found in the growing tips of plants and are detected and measured by observing the curvature which they can produce on the first leaf sheath of an oat seedling. They are called plant hormones. But much larger amounts of active substances are produced during the decomposition of proteins. These have been identified and synthesized in the laboratory and are often referred to as auxins. The first to be isolated, indole acetic acid, was called "hetero-auxin," but many similar compounds are now available and the usual test applied to estimate their activity is the initiation of root development; it is a matter of tissue differentiation or growth-regulation rather than cell elongation or growth-promotion and very small amounts of material are required. Derivatives of naphthylacetic and related compounds have been the basis of proprietary mixtures used to speed up the rooting of cuttings and also to reduce the pre-harvest fall of apples.

Of the various kinds of animal waste products used as manures, fresh poultry manure is by far the richest source of total free auxins, followed by fresh farmyard manure. Old or dried manures are much poorer but still contain important amounts. Dried blood, meat-and-bone meal and hoof meal also contain substantial quantities depending upon the degree of decomposition which has taken place before drying. Some auxins are unstable and lose their properties on being heated so that a fertilizing grade of meat meal is richer in them than a feeding grade. It is also of interest to record that peat contains a small amount.

Since the discovery of their functions, a good deal of attention has been given to the part which these compounds might play when added to the soil in a pure state. The evidence obtained has not justified the hopes that this might be a method of obtaining marked increases in crop yield. Nor has it lent support to the suggestion that the special merits of farmyard and other organic manures might be due to the presence of these substances. Excess of some of these active materials may actually retard the germination of seeds, or produce a serious physiological effect on the young plant. In fact, some of the newest specific weed killers owe their remarkable properties to the selective action of these compounds on certain weeds. This action may explain, at least in part, the commonly recognized danger to plants of fresh poultry manure.

CHAPTER XI

FERTILIZER REQUIREMENT AND THE TRACE ELEMENTS

Soil conditions—Effect of rotations—Limiting factors—Manurial requirements, chemical methods, biochemical methods, pot culture, field results—Consumption of fertilizers—Trace elements, manganese, iron, magnesium, boron, copper, zinc, sulphur, molybdenum.

THE first general theory of plant nutrition and the function of fertilizers was due to Liebig and, ignoring the early controversy over nitrogen, was to the effect that the proper fertilizer for any particular crop must contain the amounts of nitrogen, phosphorus, potassium and other elements which are withdrawn from the soil by a good yield of the plant in question. In other words, only a knowledge of the composition of the plant was needed to determine its fertilizer requirement. Farming experience and field experiments demonstrated that this was a very imperfect approximation to the truth because it failed to take into consideration the full part played by the soil. Numerous examples can be quoted to show that the best fertilizer treatments for different crops are not necessarily related to the amounts of fertilizer constituents which they take up during growth. For example, an average crop of swedes will withdraw from the soil about 150 lb. N, 30 lb. P_2O_5, and 120 lb. K_2O per acre; but the fertilizer long recognized as the most useful for swedes consists in the main of phosphate with a little nitrogen and not much potassium, if any. All plants do not exhibit such an idiosyncrasy, and the habits and conditions of growth obviously play a part; but no theory of manuring can be based on the plant alone and must take into account the capacity of the soil to provide plant nutrients.

Soil Conditions

The fertility or deficiencies of the soil can be decided only by experiment. A given soil may be able to supply sufficient phosphate for a good cereal crop but fail to meet the needs of a swede crop; or swedes may grow well on a soil on which potatoes suffer from a deficiency of potassium. The texture of the soil and the nature of its micro-flora are extremely important, for roots readily take up substances from solution only when adequately supplied with oxygen, and much of the plant's food is made available through the activity of micro-organisms. The production of plant food is also dependent upon the solvent action on the soil particles of the carbon dioxide secreted by the roots. It is not only a question of the initial concentration of plant food, but the capacity of the soil to renew the supply fast enough for the plant. Soil reactions are localized and the diffusion of soluble salts is very slow. The sharp line of demarcation to be seen between plots receiving different fertilizers is evidence that the movement of nutrients is almost wholly a vertical one. This is one reason for the common experience of the benefits to be derived from sowing a fertilizer close to the seed. Soil acidity may mean low calcium or magnesium or harmful amounts of aluminium or manganese; alkalinity may mean unavailability of iron or manganese. Soil conditions which are otherwise suitable for a crop may fail to meet its requirements under particular climate conditions sometime during the growing season, or the full absorption capacity of the plant may not fulfil the physiological demands. Hence seasonal effects, quite impossible to foretell, are not uncommon. Obviously, therefore, no single chemical criterion can assess the power of different soils to supply a specific nutrient to a particular plant.

Effect of Rotations

One explanation put forward to account for the fall in fertility of a soil was that the plant excreted toxic substances so that a rotation of crops was necessary to allow time for the decay of the toxins specific to that particular plant. The yield could then be maintained without the action of fertilizers whose chief function was to render the toxins inactive. But it is difficult to prove whether toxic substances extracted from the soil are indeed plant excretions and that they are more abundant in a soil impoverished by continuous cropping than in a fertile soil. Furthermore, a substance which is toxic to a plant in water culture may be quite harmless in the soil. But most damaging to the theory is the lack of proof that a rotation will take the place of fertilizers or that the yield will fall off only when a particular crop is grown continuously on the same land. On the rotation field at Rothamsted, the yield of wheat on the unfertilized plot slowly declined from an average of 26 bushels per acre in the first 14 courses (1848–1903) to an average of 21 bushels for the 22 courses (1848–1935). This is below the yield from the fertilized plots on the Broadbalk field which averaged 35 bushels for the first 75 years, and also below that from the fertilized plot on the same rotation field, which averaged 28 bushels although the fertilizer is applied only once in four years to the swedes which are followed by barley and either clover or bare fallow. With crops other than wheat, the fall in yield on the unfertilized plot of the rotation field compared with the fertilized plots has been greater—barley to 18 bushels from 34, clover to 25 cwt. from 52, and swedes to 16 cwt. from 300.

In these cases the rotation has been quite unable to do the work of the fertilizer—this is particularly so with the swede crop. Nevertheless, the value of a rotation of crops is undoubted in keeping the ground free of weeds and the pests associated with a particular crop, and in the successive tillage of different layers of soil; and for many crops there remains a beneficial effect which cannot be ascribed solely to the factors enumerated. Wheat and mangolds have been grown continuously at Rothamsted without any serious falling off in yield—the former for 100 years and the latter for 75—but barley has shown a distinct secular decline in yield while swedes will not grow satisfactorily for more than 10 or 12 years. So even if the possibility of a disease appropriate to certain plants is ruled out, there remains some factor which is injurious to the succeeding crop. It may be associated with the micro-flora, for partial sterilization of the soil by heat or antiseptics often brings about a marked increase in the yields of succeeding crops. The soil, however, is such a complex medium—the seat of so many diverse physical, chemical, and biological interactions—that it is essential to interpret the results of experimental work with extreme caution.

Limiting Factors

Generally speaking, the yield of a given crop is limited by the amount of one particular nutrient which happens to be deficient and, according to the well-known "law of diminishing returns," the first expenditure on the limiting factor is the most effective, each succeeding expenditure producing smaller and smaller returns. In practice this may be exemplified by the diagram of Rothamsted results (Fig. 4) in which the cost of fertilizer added to a fixed outlay of 80s. per acre is plotted with yield and cash return. The cost of production is a straight line, intersected by the curves showing returns, at points where profit ceases.

The "law" must of course be extended to all factors affecting growth for it is only when water supply and aeration

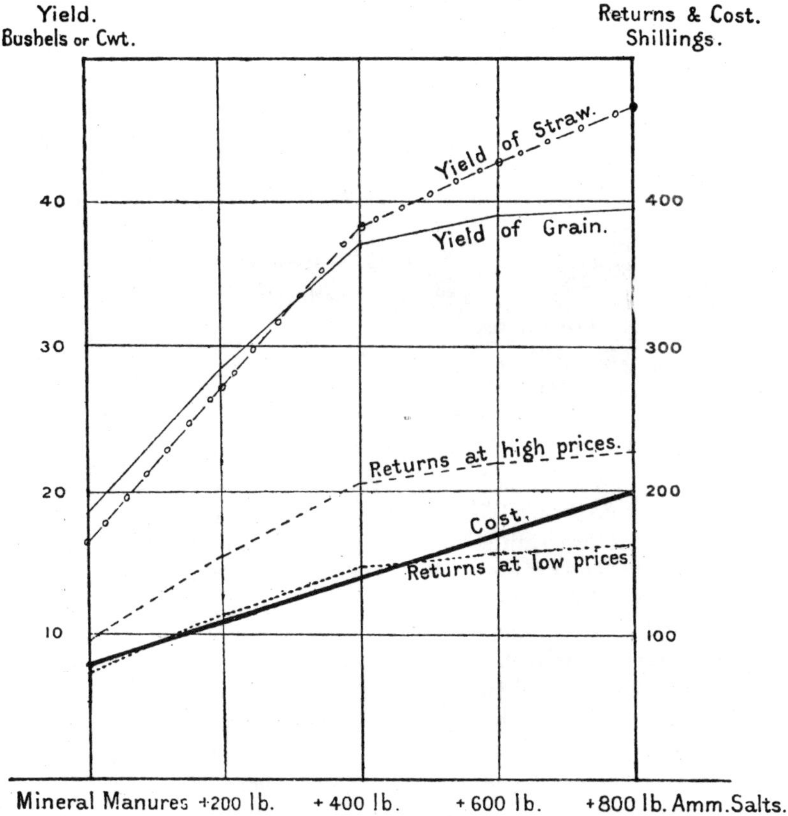

FIG. 4.—RELATION BETWEEN COST OF PRODUCTION AND RETURNS WITH VARYING QUANTITIES OF MANURE.

and temperature are adequate that the resources of the soil with respect to plant nutrients are tested so that they may become limiting factors. Fertilizers may often be wasted on land where the growth is limited by some unsatisfactory soil condition, sometimes, but not always, within the control of the farmer.

Manurial Requirements

The use of fertilizers in the past was based very largely upon trial and error demonstrations, the appearance of the crop under different treatments usually receiving greater attention than yield or quality. When one considers that even the most expert observer can seldom with confidence distinguish between two growing crops differing by less than about 20 per cent. in yield, it is obvious that important differences may easily fail to be detected by visual examination alone. Innumerable yield trials were, of course, carried out but, unfortunately, very often without due regard to the errors involved. Well-established and properly-conducted field experiments were comparatively few in number, and the conclusions reached from them necessarily applied rigidly only to the particular conditions of soil, environment and so on in question. It is obviously impracticable to carry out an unlimited number of such experiments, and much attention has therefore been devoted to studying the manurial requirements of the soil under the less satisfactory conditions obtaining in the laboratory. The problem which has confronted the investigator is the question of the availability of nutrients required by the plant. It is well known that plants may respond to additions to a soil of certain elements necessary for growth, although the soil contains large amounts of those elements. Discrimination between available and non-available constituents is very difficult, partly on account of the heterogeneity of the soil and its ever-changing character and partly because the feeding power of plants is still a matter of some doubt.

Chemical Methods.—Many purely chemical methods have been proposed and adopted. Generally speaking, they involve an extraction of the soil with water or some dilute acid, and aim at estimating the quantities of those substances which would normally come into solution under field conditions as a result of chemical and biological changes and the effects of plant growth. In certain circumstances, when the results can be properly interpreted in the light of experience, the information obtained is of the greatest value for advisory purposes; but all chemical methods must suffer from serious defects. The results depend upon the technique which is empirical, and they necessarily give only a measure of the state of the soil at a particular time, and so obviously cannot apply to its conditions before, during and after cropping; no chemical method can imitate the lengthy and complicated process of assimilation of nutrients by the plant which undoubtedly plays an active part itself in helping to bring them into assimilable form. There is probably no hard and fast line between that part of a constituent which is available to the plant and that part which is not; the compounds of the soil are continually undergoing change, and even the minerals and organic residues most resistant to decomposition are gradually broken down to supply plant food.

Biochemical Methods.—The development of certain lower organisms may form a good index of soil fertility; in other words, conditions which are satisfactory for the growth of higher plants are also usually suitable for some lower organisms. It is therefore possible to assess the cropping power of the soil by studying the development of a bacterium or fungus in the soil under suitable conditions and, since the growth of these organisms is rapid, results are obtained in a few days. For the estimation of manurial requirement the method is based essentially on the course of development of the micro-organism in a culture medium containing a definite amount of soil in place of the nutrient in question. Generally speaking, the higher plants and lower organisms react similarly to the different nutrients. Two types of organism have been employed for this purpose: (1) the bacterium *Azotobacter* and (2) the mould fungus *Aspergillus niger*.

Azotobacter is an important soil micro-organism which is able to fix atmospheric nitrogen, and its use in soil analysis originated in Denmark. It was observed, in the course of a series of liming experiments, that this organism developed when there was an adequate supply of bases in the soil but failed when the soil was acid. An adequate supply of phosphate is also necessary and the content of easily available phosphate in the soil may be estimated by studying the development of the organism under appropriate conditions. Moisture and temperature are two factors which can readily be controlled in the laboratory, and the rate of development of the organism is estimated by the number and size of the colonies of cells which become visible to the naked eye after a short period of incubation.

The Aspergillus method is similar in principle to the Azotobacter method but the necessary conditions of growth of the two organisms are quite different for Aspergillus niger develops well in acid media, and it is possible to collect the fungus and weigh it. The genus Aspergillus is characterized by the interlacing of tubular filaments to form a vegetative felt, called a mycelium, which is coherent and easily handled. The culture solution, containing all the constituents necessary for growth except that for which the soil is being tested, also contains 1 per cent. citric acid. There is, therefore, a certain similarity to chemical methods which make use of citric acid to bring easily available nutrients into solution; one difference, however, is that in the biochemical method one nutrient is being examined in presence of excess of all the others, whereas in the purely chemical methods the proportions of the nutrients present in the soil remain unaltered. As a matter of fact, the growth of Aspergillus is not so great when a citric acid extract of the soil is used in place of the soil itself. That is to say, more nutrient is withdrawn from the soil during the growth of the organism than by the extraction with citric acid. The value obtained in the Aspergillus method corresponds to the supply of nutrient which becomes available during a vegetation period lasting 5 or 6 days.

The method has no claim to being quantitatively accurate but it gives results which agree fairly closely with those obtained by other means and which serve to assess the behaviour of a large proportion of soils in the field. Other methods of assessing the available plant nutrients have relied upon the composition or yield of plants grown in various types of dish and pot experiments. Mention only need be made of two of them.

Pot Culture.—In the Neubauer method, a large number of cereal seedlings are grown in a small quantity of soil under controlled conditions of moisture and temperature so that all the root soluble or extractable nutrients are taken up in 2 or 3 weeks. The available potassium and phosphorus in the soil are determined by analysing the plants and making certain corrections for field conditions. The conditions under which the seedlings grow are highly artificial and the technique demands great skill but the method has been widely used.

The Mitscherlich method is different in principle and is designed to determine the "manurial requirement" of the soil rather than its "manurial content." It is based upon the fact that the yield of dry matter is related to the amount of any one limiting nutrient in such a manner as can be diagrammatically represented by a smooth S-shaped or sigmoid curve. It is the general and precise case of the yield curves shown in Fig. 4 and amongst the many attempts to express this law of plant growth mathematically, that by Mitscherlich is probably the best known in soil investigations. It states that the increase in yield produced by an increase in growth factor (other factors being constant) is proportional to the decrement from that maximum possible yield obtainable by increasing the particular factor.

If y_0 is the yield due to the quantity of factor already present in the soil,

y is the yield obtained by adding x of growth factor and

A is the maximum yield obtainable from this factor

then $\log (A - y_0) - \log (A - y) = cx$

where c is a constant for each growth factor.

The value A is determined experimentally in a special type of manurial experiment in which several pots in turn contain excess of all the growth factors except one; the results are then corrected for field conditions and expressed in terms of the most profitable manurial requirement.

Various criticisms of the method have been made on the applicability of the logarithmic law and the constancy of the efficiency constants as well as on the grounds of expense and time required, for a whole growing season is necessary; but it has been a great stimulation to studies on soil fertility.

Field Results.—From their study of the results of a large number of field experiments, Crowther and Yates expressed the response curve to fertilizer treatments by the equation

$$y = y_0 + d\ (1 - 10^{-kx})$$

in which d is the limiting response and k is a value assumed to be constant for each class of fertilizer and found to be reasonably close to the constants used by Mitscherlich for phosphorus and potassium. The relative responses to different dressings, the response to a standard dressing being taken as unity, are given in Table 55.

TABLE 55.—STANDARD RESPONSE CURVES FOR N, P, AND K.

N(k_N = 1·1)		P and K(k_P and k_K = 0·8)	
Cwt. N. per acre.	Relative Response.	Cwt. P_2O_5 or K_2O per acre.	Relative Response.
0·1	0·48	0·2	0·51
0·2	0·85	0·4	0·87
0·25	1·00	0·5	1·00
0·3	1·14	0·6	1·11
0·4	1·36	0·8	1·28
0·5	1·53	1·0	1·40
0·6	1·67	1·2	1·48
0·7	1·77	1·4	1·54
0·8	1·85	1·6	1·58
0·9	1·91	1·8	1·60
1·0	1·96	2·0	1·62

If the standard response curves are accepted as correct, the optimal dressings (in cwt. N, P_2O_5, and K_2O per acre) are:

For N .. .. .. $0{\cdot}12 + (\log v - \log c)/1{\cdot}1$
For P_2O_5 and K_2O .. $0{\cdot}23 + (\log v - \log c)/0{\cdot}8$

where v is the value per acre of the response to the standard dressing, and c is the cost per acre of that dressing. The optimal dressings corresponding to a series of value-cost ratios, v/c are as follows (Table 56):

TABLE 56.—OPTIMAL DRESSINGS FOR VARIOUS VALUES OF v/c.

V/c	Optimal Dressing in Cwt. Per Acre of	
	N.	P_2O_5 or K_2O.
1	0·12	0·23
1·5	0·28	0·45
2	0·39	0·61
3	0·55	0·83
4	0·67	0·98
6	0·83	1·20
8	0·94	1·36
10	1·03	1·48

The gross values of the responses to the optimal dressings are:

(a) For N .. .. .. $2{\cdot}13v - 1{\cdot}58c$
(b) For P_2O_5 and K_2O .. $1{\cdot}66v - 1{\cdot}09c$

Reference has frequently been made in previous chapters to the optimal dressings based on 1940 prices calculated from these tables.

The procedure is first to determine the value v/c by dividing the value of the increase in crop from the *standard* dressing (v) by the cost of that dressing (c); the corresponding figure for the optimal dressing is then obtained in Table 56. The gross value of the increase in crop from this optimal dressing is then calculated from the appropriate equation (a) or (b), by inserting v and c found above. The net return is obtained by subtracting the cost of this optimal dressing. The optimal or most profitable dressing is not sharply defined because there is a range of dressings over which the net return varies but slightly; the total production may of course still be raised by using more than the optimal dressing.

This outstanding contribution to the study of the best use of fertilizers is not intended to present a set of definite recommendations but rather to give figures for the relative needs of crops, which may serve as a basis for the use of fertilizers in different areas. The facts which emerge from this study by Crowther and Yates, are that the amounts of fertilizers used in this country fall considerably below the optimal dressings and that fertilizers are not always employed to the best advantage between different crops. The wartime allocation of phosphatic and potassic fertilizers to particular crops has been responsible, however, for an improvement in their most economical use for production purposes, whilst surveys of fertilizer practice now being carried out in different parts of the country will provide valuable statistics on the usual distribution of dung and fertilizers.

Consumption of Fertilizers

Meantime it is of interest to quote some of the data published by the Organization of European Economic Co-operation on the consumption of fertilizers in 1950–51. The figures in Table 57 show the amounts used on agricultural land in kilograms per hectare, which is not greatly different from pounds per acre (100 Kg. per hectare = 89·2 lb. per acre).

The outstanding feature of these statistics is the extent to which fertilizers are used in Netherlands compared with other European countries. In terms of commonly used fertilizers, the dressings correspond to about 310 lb. ammonium sulphate, 260 lb. superphosphate and 100 lb. potassium chloride per acre. The results are reflected in the fact that from 1928 to 1937 Holland produced, on the ten year average, the largest crops of wheat and sugar beet and the second largest of potatoes. The average crop production figures for the different countries correspond very closely, in fact, with the

TABLE 57.—CONSUMPTION OF FERTILIZER CONSTITUENTS (Kg. per Hectare, omitting rough grazings).

Country.	N.	P_2O_5.	K_2O.	Total.
Netherlands	71	51	66	189
Belgium	42	44	83	169
Germany	26	31	48	105
Switzerland	7	29	16	51
United Kingdom	17	34	18	70
Denmark	23	27	33	83
Norway	31	34	41	105
France	11	15	14	40
Italy	8	15	1	24
Sweden	15	23	11	49

use of fertilizer. It must be remembered, of course, that a large amount of plant food was applied to the land, especially in the British Isles, in the form of the manurial residues of imported feeding stuffs. Even allowing for the serious losses in the making and using of dung, the relatively large additions of nitrogen would reduce somewhat the wide ratio of $N : P_2O_5$ for this country compared with most other countries. Nevertheless, it is not exaggerating the position to say that enough food could not have been provided for the increasing populations without the use of fertilizers and that there is still ample scope for greater production by means of a progressive and judicious use of fertilizers.

Emphasis must always be placed on the word judicious, for both quality and quantity of crop depend upon a sensible adjustment of the plant foods to the requirements of the crop and the condition of the soil so as to build up and maintain a state of fertility. Many mistakes are committed in the use of fertilizers. One of the most common is due to an impression that old grassland contains a large store of "locked-up fertility" whereas it is often very short of lime and phosphate; these deficiencies must be attended to if the labour and cost of ploughing-out are not to be wasted. Another is the failure to realize the benefits from suitable dressings of nitrogen, especially on cereals; nor is it generally recognized that farmyard manure is a valuable source of phosphate and potassium, and its use does not reduce the crop response to inorganic nitrogen. Then there is always the tendency to apply a general mixed fertilizer irrespective of crop or soil. A greater use of the results of soil analysis would help considerably in reducing the risks of failure and the wastage of material.

There are occasions, however, when unsatisfactory crops are obtained even when careful attention has been given to fertilizer requirements. This is often due to the fact that the ordinary fertilizers are concerned only with nitrogen, phosphorus and potassium. These are the three elements usually most required to repair soil deficiencies but there are others essential to the plant although sometimes only in small amounts. These small amounts were probably present as impurities in the crude fertilizer salts formerly used, but with improvements in the manufacturing processes it is quite possible that these impurities are no longer present in sufficient quantities to compensate for natural losses from the soil. Only traces of these elements are required to ensure normal plant growth and consequently they are commonly referred to as the "trace elements."

The Trace Elements

Many of the disorders to be found in growing plants have been known for a long time and, for lack of knowledge of any casual agents such as specific organisms, have generally been attributed to "soil sickness." The "classical ten" elements: carbon, hydrogen, oxygen, nitrogen, phosphorus, sulphur, calcium, magnesium, potassium and iron were recognized as being essential for the healthy development of a plant, and although ash analyses had revealed the presence of many other elements the latter were considered to be of no importance. They were regarded as accidental, having been absorbed through the roots simply because they happened to be in the soil and the roots had not any marked powers of selective absorption. This view has been completely altered in the last twenty or thirty years, sometimes as a result of the intelligent interpretation of accidental discoveries, but latterly due to an intensive investigation of plant growth factors coupled with a great refinement in experimental technique. The result has been that several more elements have now been established as essential for plant growth. They include boron, manganese, copper, zinc and molybdenum and they are required in such minute amounts that they were at first described as the minor elements in contrast to the major elements already well known to be required in relatively large amounts. The word "minor" however suggests that they are of less importance and the term "trace element" is now more widely used. It is also common practice to discuss iron and magnesium with the trace elements, for inadequate supplies of each are often responsible for deficiency diseases and they are not considered amongst the fertilizers.

The precise function of these trace elements is not yet known; they are sometimes toxic when present in more than very small quantities; different plant species show different sensitivities to them and may display different symptoms as a result of any deficiency. Sometimes it is not a case of an absolute deficiency of the element in the soil but rather a set of soil conditions which make the element unavailable to the plant. One of the best examples of this is the manifestation of manganese deficiency when the soil contains excess calcium carbonate. Generally speaking, the analysis of leaf rather than soil has proved to be the better line of attack in confirming the reason for symptoms found in the field, but a large amount of work remains to be done to clear up the numerous difficulties which confront the investigator. The major portion of a trace element in the soil is not liberated until the minerals have undergone weathering; so it is not possible to determine the amount which will be available to the plant, without making certain assumptions regarding the ability of solvents to extract what would come into solution by natural soil processes during the growing season. It is essentially the same problem as the estimation of the available potassium or phosphate in the soil but made more difficult by the fact that in some cases the amount of trace element is extremely small.

Manganese.—It has been known since the second half of the nineteenth century that manganese was a common constituent of soils and plants and a relationship between the distribution of certain species of forest trees and their manganese content was noted in Australia early in the present century. The agricultural significance of manganese was first realized, however, when it was shown that the well-known disease of oats called grey speck could be induced by depriving the plant of manganese, and controlled by the addition of manganese compounds to the growing crop. This disease which is found in various parts of the world was particularly common on reclaimed peat soils which had been heavily limed and, as early as 1911, it was concluded that the trouble was associated with alkaline conditions especially in soils rich in organic matter.

Manganese deficiency is not related to the amount of manganese in the soil but to the form in which it exists, and to the amount which can become available to the plant. Among the many hypotheses put forward to explain the circumstances are (a) the oxidized condition and low availability of manganese under alkaline conditions, (b) the occurrence of bacteria capable of rendering manganese compounds insoluble under slightly acid or alkaline conditions and so reducing the resistance of the plant to parasitic micro-organisms, (c) the action of copper in stimulating directly or indirectly the oxidation of manganese compounds.

The characteristic symptoms of grey speck are the appearance in the middle of the older leaves of light-coloured spots which become grey, often with a reddish tint round the margin, and extend across the leaf until it collapses at the dead area. When caught in time, the trouble may be controlled by the application of a top dressing of about 25 lb. manganese sulphate per acre, or by spraying about 20 lb.

dissolved in water. Manganese deficiency also occurs in wheat and, to a lesser extent, in barley and rye.

In peas the disease called marsh spot is usually found on soils containing excess calcium carbonate and may be controlled by spraying with manganese sulphate at flowering time. The disease consists of a dark-coloured brown spotting inside the seed. Except in severe cases the foliage is unaffected. The usual symptom of manganese deficiency in potatoes is the appearance of dark brown spots along the veins of the leaves, followed in the worst cases with considerable browning of the younger leaves. It occurs mainly on alkaline soils rich in organic matter. Manganese deficiency troubles have also been found in sugar beet, garden beet and mangolds where the characteristic symptoms are a curling of the leaves so that they assume a triangular shape, and an intervenal chlorosis. An intervenal chlorosis, which may leave only the veins green in extreme cases, is the typical symptom in brassica crops. Many vegetables and both hard and soft fruits also show an intervenal chlorosis as a result of manganese deficiency.

On the other hand there are plant diseases which have been attributed to harmful effects of excess manganese. These generally occur on acid soils and include "pineapple yellows," a chlorosis in tobacco, and "crinkle leaf" in cotton.

Iron.—Although iron is one of the most abundant elements in the soil it is often unavailable in sufficient quantity to satisfy plant requirements. It is not a constituent of the chlorophyll of plants, but it is essential in the formation of this important green pigment so that the symptom of iron deficiency is a chlorosis due to lack of chlorophyll. The trouble is particularly acute amongst fruit trees and bushes, but rice and certain legumes and coniferous trees may also suffer injury from lack of iron.

The soil conditions responsible are fairly well understood. The presence of excess calcium carbonate in the soil, except when it is rich in organic matter, or a high manganese content of the soil, renders the iron unavailable. The former condition is responsible for the lime-induced chlorosis in fruit trees in this country. Large dressings of magnesium carbonate have produced the same effect. The best known case of manganese-induced chlorosis is the disease called "pineapple yellows" in Hawaii.

The deficiency is one which is difficult to control, for the soil conditions are such as to convert any iron compounds added into the insoluble ferric form. Additions of soluble ferrous salts to the soil are of little use and spraying is liable to injure the foliage. One early method of control was to encourage reducing conditions and therefore the formation of available ferrous salts around fruit trees by growing cover crops, but greater success has attended the injection of solid ferrous sulphate into the trunk of the tree and this has become almost a routine practice in some fruit areas. Lime-induced chlorosis may be caused by a lack of mobility of iron in the plant tissues due to excessive assimilation of calcium ; complex organic compounds of iron added to the soil or to sprays are claimed to correct this. Availability of iron, therefore, seems to be affected by the same factors as affect manganese availability, namely an alkaline reaction and oxidizing conditions in the soil.

Magnesium.—This is really a major element in plant nutrition and, as long as ample supplies of farmyard manure were available to add to the soil, there was little or no evidence of plants failing to obtain sufficient magnesium to produce chlorophyll of which magnesium is a constituent. A good deal of magnesium was also added in the crude potassium salts used as fertilizers. But with the modern tendency to use more concentrated fertilizers, which lack magnesium, it may become necessary to include magnesium compounds in the fertilizer programme. Magnesium is closely allied to calcium and is easily leached from sandy soils in humid climates so that the simplest remedy is to employ magnesian limestones in the routine liming operations.

The appearance of magnesium deficiency symptoms is commonly induced by excessive applications of calcium or potassium compounds. Deficiency symptoms are a chlorosis first on the older leaves and usually between the veins followed by a necrosis and premature fall of the leaf. The trouble may be the result of an actual shortage of magnesium in the soil but sometimes there are certain conditions causing inadequate absorption of the element. " Sand drown " of tobacco and cotton have been attributed to magnesium shortage and so also have certain diseases of cereals growing on acid soils. Under similar soil conditions a chlorotic and mottled effect on the lower leaves of potatoes, accompanied by a brittleness not found with the yellowing of senescence, has been controlled by moderate dressings of magnesium salts. Soil treatments have also corrected the deficiency amongst the brassicæ which are particularly susceptible, the affected leaves often developing brilliant colours. But the response to soil treatment has proved to be very slow amongst fruit and tomatoes and spraying has been more effective and much cheaper.

Boron.—The occurrence of boron in plant ash and its toxicity to certain plants was known for a long time before it was found to be not only useful but essential for normal growth. Its importance in agricultural practice became apparent with the discovery that heart rot in sugar beet and brown heart in turnips was related to a boron deficiency. The boron requirements of different species vary widely; certain members of the brassicæ are most susceptible to a deficiency, whilst cereals seem to be resistant and legumes occupy an intermediate position.

Heart rot of beet starts with the death of the youngest leaves, the disease then penetrating the flesh of the root from the crown and moving downwards. It is responsible for a serious reduction in the yield of sugar. Mangolds are also affected. Brown heart of swedes and similar crops has been reported from all over the world and consists of a browning of the flesh of the bulb with no obvious external symptom. In these cases, the trouble can be controlled by the application to the soil of about 20 lb. per acre of borax at the beginning of the season. An even distribution is important and is usually achieved by mixing the borax with the fertilizer to be applied. Other diseases which have been attributed to a deficiency of boron and have been controlled by small applications of borax are as follows: (a) browning of cauliflower which first appears as a discoloration of the stem and centre of the curd; (b) internal cork and corky core of apples characterized by the presence of dead brown spots throughout the flesh of the fruit or in the neighbourhood of the core; (c) cracked stem in celery; (d) "yellowing" in lucerne; (e) a disease of tobacco in which the death of the tip takes place and (f) a virus-like disease of potatoes.

The soil conditions responsible for boron deficiency are not yet clear in spite of a large amount of investigation. There may be a primary cause in a deficiency of boron-containing minerals in the soil, or there may be an induced deficiency, associated with alkaline conditions, especially in dry soils, and with the micro-biological activity of the soil. The addition of boron to correct a deficiency must be carried out with care for boron is required in very small quantities and excess is harmful.

Copper.—Since about 1920, "reclamation disease" has been the subject of much research in Holland, Denmark and North Germany. It occurs most commonly on acid peats which have been reclaimed and is most severe on wheat, although oats and barley are also susceptible. The usual symptoms are a loss of colour in the tips of the leaves and a serious reduction in the production of seed. In 1925 it was reported that a dressing of about 50 to 60 lb. per acre of copper sulphate provided a means of controlling the disease and, since then, many other plant diseases in other parts of the world have also been checked by the addition of copper to the soils. But it is still uncertain whether the action of the copper on the plant is direct or indirect. As far as "reclamation disease" in Europe is concerned, the soils are usually characterized by the presence of much organic matter, reducing conditions and acidity in contrast to the alkaline oxidizing conditions associated with manganese deficiency. This difference between copper and manganese may be of practical importance for there have been reports of "reclamation disease" being induced by over-correction of manganese deficiency and, conversely, the occurrence of grey speck in oats as a result of treatment with copper.

It has been suggested on the one hand that the beneficial action of the copper is due to its ability to render a toxic organic substance harmless, and on the other hand that this toxic material reduces the copper available to the plant to an amount below that required for healthy growth. Another hypothesis is that the peculiar condition of the organic matter in these soils is responsible for depriving the plants of water and that the copper sulphate coagulates the humus colloids and improves the movement of water. The characteristic symptoms can be produced on oats in water cultures containing no copper so that there seems little doubt that the plant does require a small amount of copper and is unable to obtain it under certain soil conditions, whatever the mechanism may be.

A similar disease of cereals has been met with on alkaline soils in South Australia and controlled by copper. A chlorosis of apricots, peaches and plums in South Africa, also controlled by copper, occurs on soils variable in physical properties but invariably acid. Die-back in citrus, a disease found in all citrus-growing areas, has also yielded to copper treatment whilst in certain of the organic soils of the Florida everglades spectacular responses in crop yield have been obtained by additions of copper sulphate.

Zinc.—Soluble zinc salts are toxic to plants but there are many soils, either of a sandy character in which the zinc is not present in amounts adequate to meet plant needs or alkaline in which the zinc is unavailable. There have been few reports of zinc deficiency diseases in this country but the trouble has been widespread and serious in the U.S.A. A great variety of orchard trees suffer from diseases which may usually be controlled by applications of zinc sulphate to the soil, or to the affected trees either as a spray or as an injection into the trunk. A mottling of the leaves of orange trees induced by heavy dressings of superphosphate has been recorded in South Africa. This is presumably a case of zinc being rendered immobile after absorption by the tree. A disease of tung trees called "bronzing," and a "rosetting" of pine trees have also been corrected by zinc whilst a chlorotic condition of maize, called "white bud," is prevented by zinc treatment. Zinc deficiency symptoms have also been reported in many other crops.

It has been suggested that soil conditions are not alone in causing zinc deficiency and that soil bacteria play a part in producing the symptoms. But it was not until 1931 that the curative effect of zinc was discovered and the complete function of this element in plant nutrition has still to be worked out.

Sulphur.—Sulphur is an important plant nutrient, being a constituent of the proteins, but sulphur deficiency has never been reported in this country. Nor is it to be expected, for most soils contain appreciable amounts of sulphate in solution and sulphates are added in many fertilizers—particularly superphosphate and ammonium sulphate. However, on a highly leached and impoverished soil of Nyasaland a deficiency of sulphur has been found in tea, so called "tea yellows."

Molybdenum.—This element is essential for nitrogen fixation and metabolism and for the healthy growth of clover. A deficiency of available molybdenum gives rise to "whiptail" in cauliflower—usually on acid soils.

Sufficient has been said to show the widespread effects which a very small amount of some elements may have, sometimes beneficial, sometimes harmful. Modern work, aided by the development of micro-chemical and spectrographic methods of analysis, are bringing many curious facts to light. But much work still remains to be done before any complete picture can be built up to show the exact function of all these trace elements in the plant. Furthermore, it is important to remember that a very large number of factors are concerned in the development of living organisms. It is not to be expected that the fairly precise symptoms induced and cured under carefully controlled experimental conditions will be so easy to diagnose in the field. The possible antagonism of two or more elements must constantly be borne in mind and so also must the influence of environment and the possible damage from fungi, insects and various harmful viruses. Visual observation is fraught with many difficulties in those respects and in many cases a reliable diagnosis can be made only by the experienced investigator who has laboratory facilities to confirm his conclusions.

CHAPTER XII

SYSTEMS OF MANURING CROPS

General considerations—Cereals, wheat, barley, oats, rye, maize—Root crops, swedes, other crucifers, mangolds, potatoes—Leguminous crops, beans, clover, lucerne, sainfoin, vetches, trefoil—Grassland, hay, pasture—Hops—Fruit plantations—Tropical and sub-tropical crops, sugar cane, cotton, tobacco, tea—Gardens.

General Considerations

IN dealing with the specific properties of the various fertilizers, a number of illustrations have been given from the results of field experiments on particular crops, from which conclusions might be drawn as to the fertilizers most appropriate to those crops, but in the main these experiments have been selected to illustrate the action of the fertilizer rather than the requirements of the plant. It remains to reconsider the information derived from experiments under its practical aspect, so as to obtain a guide to the methods of manuring which the farmer should adopt for the crops he is setting out to grow. It is never possible to do this absolutely; the proper manure for any particular crop must always be conditioned by a number of local circumstances special to the farm in question; from which it follows that the mixtures sold as "Turnip

Manures," "Potato Manures," and so forth, must often be more or less wasteful if they are to be effective everywhere. Instead of applying a kind of average manure, the farmer ought to have such an appreciation of manurial principles that he can adapt his fertilizers as economically as possible to his own soil and conditions of farming.

In discussing the application of fertilizers to crops, even when the special features presented by the soil are neglected, we can draw no conclusions as to the proper methods of manuring unless we take into account the place the crop occupies in the rotation adopted by the farmer, and also the character of his land and style of farming. For example, we have not to consider the wheat crop as standing by itself in the manner we see it in the Rothamsted experiments, but as it is generally grown in practice—after a clover crop, or perhaps after mangolds or potatoes which have been manured with dung. Furthermore, one man may be in possession of good land in high condition, and may be farming "high" for big crops; he will be justified in a greater outlay upon fertilizers than would be advisable for an equally good farmer on poorer land, where it may be more economical to be content with smaller crops and to keep down the expenditure. The manuring to be adopted on a given farm must be looked at as a whole, as a system to be shaped as much by various wide considerations of farming policy as by the particular crops that are being grown. It is easy, for example, to indicate the composition a manure for swedes should possess, but whether a farmer should spend 20s. or 60s. an acre on fertilizers for his swede land depends entirely upon the general character and style of his farm. The amount of fertilizers that can be profitably employed, will always be dictated by such local conditions as the markets available, the supply of labour, and the rent of the land.

On the one hand we have had the systems prevailing in the middle west of America and other more newly settled countries where the farmer lived upon the capital originally stored up in the virgin soil. He grew, for example, maize and wheat alternately, using no fertilizer and restoring nothing to the soil, often burning the straw and not even taking the trouble to cart out the manure accumulated beneath any cattle he might feed. Year after year 50 to 100 lb. of nitrogen per acre were removed and the soil became steadily poorer, yet it proved to be more profitable to move to fresh land than to spend money in restoring the lost fertility. On the other hand, in many parts of Great Britain we may see a strictly conservative system at work. The land possesses a certain condition and will yield fair average crops, only part of which are sold, the rest being converted into meat, milk and dung, by which means the greater part of the plant food drawn from the soil is returned. There is, however, a certain removal in the corn and meat and a certain amount of waste in dung-making, but this is repaired by the growth of clover, and by the purchase of a comparatively limited amount of fertilizers or feeding stuffs, so that the condition of the land is maintained but not at a very high level. Again, at the other end of the scale we have the intensive farmer who uses his land as a sort of manufacturing medium to convert fertilizers into crops, and steadily increases the fertility of his soil by adding much more plant food than he removes in his crops.

We can begin by considering what is necessary to maintain the condition of the land under a conservative system of farming, and we may take the case of a farm under a four-course rotation, where nothing but corn and meat are sold and all the dung goes back to the land. Under such conditions as we have already learnt, the feeding animals retain only about 10 per cent. of the fertilizing constituents of the food they consume; the other 90 per cent. comes back in the manure and wholly or in part reaches the land again (Table 58).

In this way the land loses 22·6 lb. of nitrogen per acre per annum; but this estimate fails to take into account the very considerable losses that occur during the making of the farmyard manure, which may be estimated at 50 lb. in the four years, and those due to drainage and bacterial action. On the other hand, the nitrogen contained in the clover crop has

TABLE 58.—FERTILIZING CONSTITUENTS REMOVED FROM FARM IN CORN AND MEAT SOLD.

	Nitrogen.	P_2O_5.	K_2O.
	Lb.	Lb.	Lb.
Swede Turnips, 20 tons fed . .	10·0	5·8	0·7
Barley, 6 quarters sold . .	40·0	18·0	12·0
½ ton Straw fed . . .	1·2	0·7	0·1
The rest made into Dung.			
Clover, 2 tons Hay fed . . .	4·1	2·8	0·4
Wheat, 4 quarters sold . . .	35·2	15·3	10·3
Straw made into Dung.			
Total for 4 acres . .	90·5	42·6	23·5
Per acre per annum .	22·6	10·6	5·9

been obtained from the atmosphere; indeed, the Rothamsted experiments have shown that the land is left richer in nitrogen after a big clover crop has been grown and taken away.

A further consideration of the rotation field at Rothamsted shows that the clover crop alone would be able to maintain the fertility of the land at about the condition which would produce such yields as are shown in the table. For instance, the Agdell field in the 47th to the 50th years gave the following crops on the portion which had received no nitrogen throughout the whole period, though phosphates and potassium were supplied to the swede crop.

If, then, in this case the swede turnips had also received whatever manure would have been made from the clover

TABLE 59.—PRODUCE OF AGDELL FIELD UNDER ROTATION. NO NITROGEN SUPPLIED IN MANURE. (Rothamsted.)

1894	Clover Hay	Carted away . .	64·7 cwt.
1895	Wheat .	Carted away . .	39·6 bushels, and 25·3 cwt. Straw.
1896	Swedes .	Consumed on the land	12·0 tons.
1897	Barley .	Carted away . .	37·7 bushels, and 24·9 cwt. Straw.

hay and the wheat and barley straw, it is evident that the production would have been little short of the average indicated in Table 58, and that the nitrogen necessary to maintain the fertility of the land at such a level would be supplied indefinitely by the recurring clover crop. In the Agdell example phosphatic and potassic fertilizers were, however, freely employed, and it is obvious that the soil possesses no power of increasing its stock of these constituents in the same way as it can obtain nitrogen from the atmosphere. Three hundred pounds of superphosphate per acre during the four-year period of rotation would, however, repair the losses, and the losses of potassium are so small that on a loamy or clay soil they would be made up by the continual slow weathering of potassium compounds in the soil.

It is, however, a low level of production that is attained in this example of an almost self-supporting piece of land, and if the average yield is to be raised, say to 5 qrs. of wheat and 6 qrs. of barley per acre, an external supply of nitrogen must be obtained, either in the form of fertilizers or feeding stuffs. Moreover, this additional nitrogen must be considerably more than would be contained in the extra quarter of wheat and other larger crops that are grown; there must be enough to compensate for the greatly increased waste by drainage, denitrification, etc. In short, it must be remembered that as the fertility of the land is increased all the soil processes are speeded up and the plant may not be able to use all the nutrients which become available. Hence, the intensive farmer often becomes wasteful because, after his land is in good heart, he continues to add fertilizers at the same rate as he did when he was building up its condition.

It therefore follows that an account of what is removed from the soil year by year by the crops or animals raised upon the farm may not provide adequate guidance as to the amount of fertilizer which must be brought in; at a low level of production, good land will practically recuperate itself without any extraneous manure, while high farming for big crops will involve some loss of the nitrogen applied to the land but never recovered in the crop. It is only by experience, by the knowledge of his own land and the market conditions which prevail, that the individual farmer can tell how high it is profitable for him to farm, and therefore to what degree he can utilize the information as to feeding his crops which is provided by field experiments.

The discussion that follows of the manures appropriate to each of the staple crops is therefore simply intended to help the farmer to build up a rational system appropriate to his own farm. In the practice of farming, many things may at once be set down as "wrong," but there can be nothing absolutely "right"; the proper course of action is never anything more than a judicious compromise adapted to all the various conditions of climate, soil, and markets. We can now consider the ordinary farm crops separately.

Cereals

Wheat in the typical four-course rotation follows the ploughed-up clover ley, and generally derives all the nitrogen it requires from the residues left by the clover in the soil. In many cases, however, oats are now substituted for wheat after the ley, because more time is thus obtained to graze the aftermath and break up the land before seeding; further, after the ploughed land has been exposed for the winter, oats suffer less than wheat from the wireworm which is apt to be prevalent in the old clover land. Should wheat follow a good crop of clover, further manuring is not required; though if the second growth of the clover has been allowed to ripen seed, which removes a large proportion of the nitrogen, or if much rye grass has been present in the seeds mixture and the clover has failed somewhat, it may be desirable to enrich the ground still further. This may be done either by spreading a coating of dung (10 tons per acre) on the clover before ploughing, or by a spring top-dressing of 1 to 1½ cwt. per acre of one of the nitrates or ammonium sulphate. The same applies when wheat follows mangolds or potatoes which have received dung and fertilizers. Speaking generally, on soils in good heart it is wise to wait until the early spring; if a wet winter has washed out the soil nitrates and the plant appears to be growing badly or losing ground, a top dressing of nitrogen will be particularly beneficial. Soot has for some centuries been employed as a spring top-dressing for wheat; besides the nitrogen it supplies, it also tends to preserve the plant from the attacks of the small slugs and snails which are so active at that time of year.

Of course, when wheat and other cereals are grown continuously on the same land it is necessary to employ a more complete fertilizer—2 cwt. per acre of nitrogenous fertilizer will be required as a spring top-dressing, and 3 cwt. of superphosphate or 2 cwt. of basic slag, according to the condition of the soil, should be sown before the seed. Potassium is necessary only on the lighter soils, on which wheat is not likely to be grown continuously, but in such a case about 1 cwt. per acre of potash salt would be desirable.

Barley is grown under two very different conditions of tilth. In the first place, it may follow wheat and form the second or even the third white straw crop after roots or a clover ley; in the Isle of Thanet three, four, or even five barley crops may be taken in succession after an old lucerne or sainfoin ley has been broken up. In such cases the high condition will have been taken out of the soil by the first crop of wheat, there will no longer be any excess of readily available nitrogen, and as there is a good opportunity of getting the soil early into tilth, barley of high quality may be expected. Good malting barley contains a relatively low percentage of nitrogen, hence the soil on which it grows must not be too rich, nor must any large quantity of nitrogenous manure be employed. On the other hand, however, it is a mistake to suppose that impoverished soil alone will yield good barley; unless a reasonable amount of nitrogen be available not only will the yield be small but the size of the berry will fall away. In preparing for a crop of barley of high quality it is therefore necessary not to allow the land to become really poor, and it is desirable that the nitrogen applied should balance the available phosphate and potassium. If the land is in really high condition before the first straw crop of wheat or oats is taken, barley may follow without any fertilizer, especially if the ground can be got into good tilth and the barley sown really early. But for a second barley crop or for the first on land in poorer heart some nitrogenous manure must be used, and ammonium sulphate and rape cake are found to give better quality than sodium nitrate, though in neither case must a large quantity be used. The crop will also benefit from a dressing of about 2 cwt. per acre of superphosphate, or its equivalent in some other form of readily soluble phosphatic fertilizer. The question of potassium is more doubtful; while potassic manures have been found to stiffen the straw and increase the size of the berry by promoting starch-formation, they also prolong the maturity of the barley and darken its colour slightly. But they are likely to be valuable on light sandy or chalk soils. We thus arrive at the following mixture for a barley manure, when barley follows one or more white straw crops and the land is no longer in high condition:

Ammonium sulphate ½ to 1½ cwt., or rape dust 4 to 6 cwt. per acre; superphosphate 2 cwt. per acre; potash salt 1 cwt. per acre where required.

The fertilizers should be mixed and sown before the seed is drilled; a top dressing of ammonium sulphate is usually of little value. A mixture of this kind would also serve for the rare case of barley following roots which have been grown without farmyard manure and then carted off the land.

When barley follows roots which have been highly manured with farmyard manure, still more so when the roots have been folded off by sheep, the land is already too rich to grow barley of the highest quality, the more so as the roots are often left so late on the ground that a good seed-bed cannot be obtained early in the year. Early sowing is essential for barley of high quality and when this is not possible it is better to sow either oats or a barley like Archer's, which will yield well for feeding purposes though the quality may not reach a malting standard. When the roots have been fed on the land, 3 cwt. per acre of superphosphate sown with the seed is found to improve the quality of the grain and helps to correct the excess of nitrogen; but potassic fertilizers are of little value except on very light or chalk soils. When the roots have been grown with farmyard manure and phosphates and then carted off, the land will be in about the right condition for barley, and will want only a light dressing of nitrogen.

Oats.—The general principles of manuring for barley hold also for oats, except that, being grown for feeding purposes only, they can be given much larger quantities of nitrogen without any fear of injuring their quality. When grown on a ploughed-up ley, which in many cases is also lightly dunged before ploughing, oats are not likely to require any fertilizer; at the most a little sodium nitrate if they are found to be starting away too slowly. As an all-round fertilizer for oats when the land is in poor condition 1 to 2 cwt. of sodium nitrate or ammonium sulphate, and 2 cwt. of superphosphate or basic slag, according to the class of soil, will answer all the requirements of the oat crop; potassic fertilizers would be wasted, as also would the more expensive organic forms of nitrogen with a crop which occupies the land for so short a period. Of course, in a wet district 2 cwt. per acre of nitrogenous manure might easily result in the lodging of the crop.

Rye, which is grown in the south of England for early spring keep is rarely manured; but *maize*, which is also grown to some extent as fodder, requires the land to be brought into fairly high condition. A preliminary dressing of 12 to 15 loads of dung per acre should be given, with 2 to 3 cwt. per acre of superphosphate at the time of sowing, then 1 cwt. per acre of sodium nitrate may be used as a top dressing round the plants when they are set out and side hoed.

Root-crops

In British farming the bulk of the manure that is made upon the farm or purchased is applied to the root-crops—swedes or mangolds; though in the east and south-east of England it is more general to apply the farmyard manure to the seeds before ploughing up for wheat. In these warm soils much nitrogenous manure is apt to cause swedes to run to top and to be more susceptible to mildew. Big crops of roots mean more food for the stock, and so in turn more farmyard manure. Moreover, the roots are grateful, and continue to respond to liberal treatment without lodging or growing an excess of straw, as cereal crops will do. It is questionable, however, whether the very common practice in the north of putting on all the available manure, farmyard and artificial, for the root-crop and making that serve for the whole of the rotation, is wise; better results will be obtained by a careful adaptation of the fertilizer to the particular crops forming the rotation.

Swedes.—The earliest work that was done at Rothamsted consisted in showing the dependence of this crop upon an ample supply of phosphatic manure of an available character, and it was the response of this crop to soluble phosphates which largely encouraged the superphosphate and other fertilizer industries. The point may be illustrated from the Rothamsted experiments on the Agdell field, where crops are grown in rotation with the following average results:

Unmanured	16 cwt.
Minerals only—Superphosphate and potassium sulphate	208 ,,
Complete Manure—Nitrogen, Superphosphate, and Potassium Sulphate	400 ,,

Without manure the yield is trifling, but with the mineral manures (and the phosphate is the effective factor) the yield rises to 208 cwt. per acre, although the land had been continually cropped without any nitrogen supply; lastly, when nitrogen also is added, the yield becomes that of a high average crop for the south of England. In practice, however, it is found that where the land has been kept in good condition and there has been adequate preparation of the seed-bed, little manurial nitrogen will be required to supplement the nitrates produced from the soil reserves, and that consequently the great increase due to the nitrogen in the experiments quoted will not be reproduced under ordinary conditions of farming.

In a large co-operative series of trials undertaken by the Highland and Agricultural Society over the whole of Scotland it was found that 84 lb. per acre of ammonium sulphate, or its equivalent in 1 cwt. of sodium nitrate, was as much nitrogenous manure as could be profitably employed. About 5 cwt. per acre of superphosphate, or 4 cwt. of basic slag, or 2 cwt. of steamed bone flour, according to the soil, were indispensable; the superphosphate being best on loams and calcareous soils, the basic slag on clays and peaty land, and the steamed bone flour on sands and gravels.

Phosphate alone put up the crop from 11·3 to 17·9 tons per acre: 1 cwt. of sodium nitrate or 84 lb. ammonium sulphate added another ton to the crop, while 2 cwt. nitrate produced no additional increase. See p. 61 for a wider range.

The question of the most appropriate manurial treatment for swedes depends upon how much farmyard manure is available; while the ordinary four-course rotation is being practised, most of the dung made will come back to the land for the swede crop, about 10 tons to the acre being available. With such quantities of farmyard manure the swedes will require little further nitrogenous dressing; phosphates are, however, still indispensable. In such cases it is generally the custom to finish off the seed-bed preparation with a ridging plough, and to apply the dung to the furrows just before sowing. The ridges are then split back over the dung, the new ridges thus formed are rolled, and the seed and superphosphate are sown from the same drill on the top of the ridge. This plan answers excellently in the cooler and moister parts of the country, where the swede flourishes and grows big crops, but in the south and east of England such a method exposes the crop too much to risk of damage from drought, both through evaporation from the sides of the ridge and because the fresh manure as it rots leaves the land too open. On warm dry soils it is better to plough in the farmyard manure in the autumn, and to sow the swedes on the flat with their appropriate fertilizer. It is in the south again that farmyard manure is often lacking for the swede crop, because it has been wanted for wheat or hops or potatoes, or sometimes for the grassland; many sheep farmers, who fold on the swede land, have a strong objection to swedes grown with farmyard manure. A suitable mixture, when no farmyard manure is available, consists of 4 cwt. of superphosphate (or its equivalent in basic slag with a high solubility), 2 cwt. of fish or meat guano, and ½ to 1 cwt. of nitrogenous fertilizer as a top dressing when the plants are singled. If the land is in really good heart, the fish guano can be omitted or reduced. It will be seen that various compounds of nitrogen are used in order to ensure a steady and continuous supply of nitrates as long as the plant is growing. Though superphosphate and ammonium sulphate are, on the whole, the best fertilizers in their respective classes for swedes, they must be employed with care where the soil is acid, and especially if the land is known to be subject to "finger-and-toe." The organism causing finger-and-toe flourishes in an acid medium, so that lime should be applied or superphosphate replaced by basic slag. Potassium salts are sometimes omitted for the swede crop, though, like other root-crops storing up a good deal of carbohydrate, the swede will respond to liberal allowances of potassium. On the lighter soils, when farmyard manure is only scantily used, it is undoubtedly wise to apply about 1 cwt. of potassium chloride while the land is being prepared for the seed-bed.

Of the other crops allied to swedes, white turnips require much the same treatment, except that the fish guano may be omitted because turnips possess a shorter period of growth, while the potassium is more necessary. Kohl rabi may have just the same treatment as swedes, as may thousand-headed kale and cabbage, with the addition of more nitrogen. Cabbages and kale in particular will respond to enormous quantities of nitrogen; in addition to the farmyard manure or fish guano recommended for the swedes, up to 5 cwt. per acre of nitrogenous fertilizer may be used in two or three top dressings.

In market garden work such active nitrogenous manure brings the cabbages earlier to cutting and renders them more tender, though they are reputed in consequence not to travel so well to distant markets. Stock feeders, as a rule, do not like cabbages or any root-crop grown with an excessive amount of nitrogen, on the ground that it becomes a poor or even a harmful food; but whether this is due to the increased amount of nitrates in the plant or to other compounds of nitrogen is as yet uncertain.

Mangolds are often described as heavy feeders, by which we may understand that the yield will go on responding to very large additions of manure rather than that the crop removes a specially large amount of manurial constituents from the soil. The mangold differs from the swede in its requirements. It gives returns for large quantities of nitrogen; it needs potassium but not much phosphate. The Rothamsted experiments show that mangolds can be grown successfully for very many years in succession upon the same land if suitable fertilizers are provided. The only difficulty experienced lies in the getting of a plant on the plots where the tilth of the soil has been injured by long-continued treatment in one particular direction. The results given by some of the plots are set out in Table 60.

TABLE 60.—AVERAGE YIELD IN TONS OF MANGOLDS (Rothamsted). 32 YEARS, 1876–1907.

	Superphosphate.		Dung.	
	No Potash.	Potash, etc.	Alone.	With Phosphate and Potash.
Rape Cake = 98 lb. N.	11·1	22·0	24·5	25·7
Nitrate of Soda = 86 lb. N. . .	15·3	18·0	25·9	26·4
Ammonium Salts = 86 lb. N. . .	7·5	15·2	22·5	24·0

These results illustrate the following points in the manuring of mangolds:

(1) The value of dung and of organic manures like rape cake, which, by maintaining a good texture in the soil, ensure a plant and a vigorous start.

(2) The value of an addition of active nitrogenous manures, particularly sodium nitrate, even when dung is also used.

(3) The importance of potassium salts even when dung is also used. The beneficial effect of potassium salts is less apparent when sodium nitrate is employed as a source of nitrogen, because the sodium is also a nutrient. Potassium thus becomes more necessary when ammonium salts or rape cake form the source of nitrogen; but it is desirable to use some sodium salt, such as common salt itself, as an economizer of the more valuable potassium.

(4) That with proper manuring, mangolds can be grown year after year on the same land without any falling-off in yield or any accumulation of disease. It is sometimes convenient to keep a little piece of land near the homestead always in mangolds; this can be done for a long time in perfect safety if organic manures are employed to maintain the texture of the soil.

Coming now to the requirements of the crop in practice, not much variation will be required because of its position in the rotation, since mangolds are practically always grown on a stubble with the land in comparatively poor condition. The basis of a manure for mangolds should be dung to which a large response is obtained. On light soils and in dry climates it is better to plough in the dung in the autumn and grow the mangolds on the flat, lest the fresh manure should leave the soil too open and let in the drought, but on heavier land and where the rainfall is greater the land will generally be laid up in ridges. The dung should be spread in the furrows; the fertilizers, other than sodium nitrate or other active nitrogenous manure, should be sown on the dung and the ridges then split back on to the dung. Supposing 20 loads of dung per acre to be available for the crop, the supplementary manure should consist of 3 to 5 cwt. per acre of potash salt (the larger quantity on light soils), and 2 cwt. of fish guano or kindred fertilizer if the land is in poor heart and a large yield wanted. Phosphates in many cases, as at Rothamsted, are not required when dung is used, but on soils where phosphates are specially necessary, as on many of the clay soils so suited to the mangold crop, it will be well to add 2 cwt. of superphosphate to the mixture whenever the fish guano is omitted. The after-treatment will consist in giving top dressings of a mixture of equal weights of sodium nitrate and salt; about 3 cwt. of the mixture at singling time, and perhaps an equal amount a few weeks later, should a specially heavy yield be aimed at.

Sugar Beet.—The growth of sugar beet is now of great importance to British farming, especially on the lighter and warmer soils. The labour involved in lifting, cleaning, and harvesting beet on the strong soils adds excessively to the cost of production whilst it is only on the warmer soils that the crop matures sufficiently rapidly and attains its proper percentage of sugar.

The manuring of sugar beet follows the same general lines as that of mangolds, as would be expected from the fact that both plants are selections for special purposes from the same original stock. The application of fertilizers for sugar beet may, however, be rather heavier in view of the greater value of the crop; fresh dung has also to be avoided since it leads to the production of forked and unshapely roots. Indeed better sugar beet is grown upon land that is already in high condition than upon poor land that has to be heavily enriched for that particular crop.

The best preparation is a dressing of dung, at least 10 and up to 15 tons to the acre, applied in late autumn and ploughed in. In the spring a mixture should be prepared of 3 cwt. superphosphate, 3 to 4 cwt. potash salts or 2 cwt. potassium chloride, and 3 cwt. ammonium sulphate, and this should be applied immediately before sowing the beet. It should be broadcast by machine and harrowed in or, if the land is laid up in ridges, it can be sown in the furrows where the seed is to follow and covered by splitting the ridges. Salt at the rate of 3 to 5 cwt. per acre, sown sometime before seeding, may be used instead of the potassium salts. When the sugar beet is singled, a top dressing of 1 cwt. per acre of sodium nitrate or of nitrochalk should be applied. A remunerative return is often obtained for a second top dressing applied three weeks or a month later, before the leaves begin to meet in the rows. Whether this second dressing is needed or not must be decided by the condition of the land, the season, and whether the crop is growing vigorously or not.

In many cases the requisite amount of dung will not be available and it is desirable to employ some slow-acting form of nitrogen in its place. Where shoddy is accessible the methods of the hop grower or fruit grower may be adopted and 2 tons of shoddy per acre may be ploughed in during the winter in place of the dung, to be followed by the other dressings stated above. As an alternative, the mixture to be applied before seeding should consist of 10 cwt. per acre of fish guano or its equivalent containing about 7 per cent. of nitrogen, 2 cwt. of superphosphate, 3 cwt. of potassium chloride and 1 cwt. of ammonium sulphate, to be followed by the top dressings as before. This mixture should be sown by machine or broadcast and worked into the soil a week or so before drilling. When the land is in good heart satisfactory crops can be grown by increasing the dressings of inorganic fertilizers alone. The effectiveness of the different fertilizers depends a lot upon the soil and the season, but nitrogen is most consistently valuable, with or without dung.

Potatoes.—It is more than usually difficult to lay down general rules for the manuring of the potato crop, so varied are the tilths upon which it is grown and so different are the yields that are aimed at. In the neighbourhood of great cities where dung was cheaply obtained the farmer would often crop suitable land every other year with potatoes, taking a cereal or a green crop in the intervening years. In Ayrshire, some fields have grown early potatoes continuously for 40 years or more, usually being followed by a catch crop eaten off by sheep. On the other hand the farmer who does not make a speciality of potatoes will simply plant them on a portion of his mangold or swede land, while in good potato-growing districts they will form one item in a five- or six-year rotation. In the Lothians, for example, a common rotation is roots, barley, hay, oats, potatoes, wheat, about 15 tons of farmyard manure being put on the stubble and ploughed in the autumn before the potatoes are grown, and fertilizers to the extent of 10 or 12 cwt. being added in the spring. A rotation in the Dunbar area, so famous for the high quality of its potatoes, is as follows: Swedes, in part fed on the land; barley; clover, cut for hay; clover, grazed with cake and corn; potatoes, no farmyard manure; oats.

On the Lincolnshire silt soils a common rotation is: Swedes ½, potatoes ½, with farmyard manure; wheat; seeds; wheat; oats.

On the black soils of Lancashire a common rotation is: oats; potatoes; oats; seeds; farmyard manure being used in large quantities.

In view of such variations in practice it will be best to discuss a few general principles. Potatoes should not receive an excess of nitrogenous manure, because it affects their quality and makes them susceptible to disease. As quality is so important, the nitrogen should be carefully balanced by potassium and phosphate. A good supply of phosphatic manure has been shown to be important and potassium is essential since the potato is a starch-making plant. Ammonium sulphate is better than sodium nitrate and superphosphate than basic slag; lime also should be used sparingly if required.

As regards the use of dung, it has been repeatedly shown that a better return is obtained by using farmyard manure in moderate quantities of 20 loads per acre or so and supplemented with fertilizers, than by spending all the money available for manuring upon dung alone. On any but the heaviest soils it is better to plough in the farmyard manure in the autumn and so get the land into good heart; on close badly working soils, it is an advantage to the potato plant to have the ground left a little hollow by the decay of the farmyard manure so the dung should be applied in the drills just before planting. The mixture of fertilizers should either be sown broadcast after the land is ridged up or sown upon the farmyard manure in the drills before the ridges are split. For ordinary cropping a mixture of 3 to 4 cwt. per acre of superphosphate, 2 cwt. of potassium sulphate, and 2 to 3 cwt. of ammonium sulphate will be ample; when extra heavy crops are aimed at, the dressing should be increased.

The Leguminous Crops.—It has already been explained that the leguminous plants are able to obtain nitrogen from the atmosphere by the agency of the bacteria in their nodules and can in this way satisfy their requirements for nitrogen; it should not be forgotten, however, that they also feed upon combined nitrogen like all other plants, and so derive their nitrogen from the air and from the soil. To obtain the biggest crops, a rich soil and some nitrogenous manure are necessary, but to secure the greatest profit out of a leguminous crop, it should be left as far as possible to derive its nitrogen from the atmosphere. All leguminous plants are sensitive to soil acidity and need calcium. Hence special attention should be given to the lime status of the soil.

Beans.—Beans no longer play the important part in British agriculture that they once did; essentially a heavy land crop, the cultivation declined when much of the strong clay land was laid down to grass. In the rotation, beans generally come between two white straw crops. They will follow oats or barley, for example, and precede wheat, and as a rule they do not receive any manure. A little farmyard manure may be spread on the stubble before it is ploughed, but other nitrogenous manures have little beneficial effect upon the crop. The Rothamsted experiments show that beans, like other leguminous plants, respond chiefly to phosphates and potassium, to the latter especially, and are able to derive most of the nitrogen they require from the atmosphere.

Experiments by the Highland and Agricultural Society in Scotland, and others in Essex, under ordinary farming conditions confirm these results. Thus in practice, when beans are being grown on strong land, we may reduce the manuring to 3 or 4 cwt. per acre of basic slag; any other expenditure on fertilizer is most likely to be repaid by an increase in the use of dung.

Clover.—Red Clover forms one of the most important crops cultivated by the farmer; not only does the hay furnish a particularly valuable fodder, the nitrogen in which is largely derived from the atmosphere and is therefore clear gain to the farm, but the nitrogen left behind in the roots and stubble also enriches the land for future crops.

Since the time of the Romans it has been known that the wheat is most luxuriant where the clover had grown best in the preceding year; the Rothamsted experiments afford some interesting examples from which the gain of nitrogen can be estimated. One example of the great benefit which the succeeding crops in a rotation derive from a good crop of clover, although it is removed from the land as hay, has already been quoted (Table 6). Again, in 1873 a piece of land in Little Hoos field was cropped, part with barley and part with clover; in 1874 barley was taken over the whole. The amount of nitrogen removed in the crop from each piece of land was 151 lb. per acre in 1873 from the clover portion of the field, as compared with 37 lb. from the barley portion, yet the former in the following year yielded an extra 20 bushels of barley.

As to the manurial treatment of clover, it is difficult to quote very extensive experiments, because of the failure of the plant which takes place through clover "sickness." On some of the best clover soils in this country it cannot be grown more frequently than once in four years; sometimes more than once in seven or eight years is not safe. (See Chapter II.) Manuring alone will not keep off clover sickness, though lime and potassium salts are helpful. In practice clover is rarely manured; it is nearly always sown in the barley crop, and is then left to the mineral residues from the preceding root crop and the nitrogen it can gain from the atmosphere; at the most, a little farmyard manure may be spread during the winter and is valuable as affording shelter to the young plants. If plenty of phosphates have been used for the swede and barley crops, nothing more in this direction is likely to be required, but on many soils, especially of the lighter kind, an application of potassium during the late autumn or winter after the clover has been sown will have a marked effect upon the yield of clover, and the cost of about 1 cwt. of potassium chloride per acre will be amply repaid.

It is rarely wise to attempt to manure standing clover for a second year's crop; nitrogenous fertilizers are not required, and the potassium and phosphates hardly have time to get well down to the plants' roots in the time the crop still occupies the ground. A thin coating of dung in the winter is valuable for its shelter, and if the crop must be encouraged, then 4 cwt. of a potassic phosphate mixture may be sown broadcast in the winter; even if it does not produce much immediate return, it will not be washed away.

Lucerne and Sainfoin.—The principles which have been laid down for the treatment of clover apply equally to lucerne and sainfoin (i.e. that mineral manures should be used, and that only the young plant will respond to fertilizers), but since these crops are generally sown to stand five years or more, it is wise to make a good preparation of the soil before sowing. As a rule, they are sown in barley or oats and about 5 cwt. per acre of basic slag should be worked into the soil before sowing the corn crop. The potassium chloride (2 cwt. per acre), being soluble, can be kept until the autumn or winter. Beyond this it is not wise to use fertilizers on these crops; a little sodium nitrate may serve to give the young plant a start in its first spring, and a coat of dung is often valuable, but the proper way to regard lucerne or sainfoin is as a cheap means of enriching the land with a minimum of expenditure.

Vetches, Trefoil, Crimson Clover, and similar rapidly growing leguminous crops are usually grown as catch crops on land that is already in good heart and they do not require fertilizer. Lupins are sometimes grown on poor sandy land in order to be ploughed in as green manure; in such a case the preparation of the land (supposing it to be poor healthy land undergoing reclamation) should include the application of 4 to 5 cwt. per acre of basic slag and 1 to 2 cwt. of muriate to supply the lupins with the necessary mineral food, without which they could neither gather nitrogen nor accumulate humus for the amelioration of the soil.

Grass Land

In considering the effect of manures upon the grass crop, we have to take into account not only the weight of the produce but the character and botanical composition of the herbage that ensues. Every meadow possesses a characteristic vegetation made up of various species of grasses, a few leguminous plants like white and red clover, bird's foot trefoil, the yellow vetchling, etc., and sundry miscellaneous species which are, in the main, of little value to stock and may be classed as weeds. The proportion which each of these species contributes to the herbage represents the degree to which it is suited by the various conditions of food, water, soil, texture, etc., which prevail in that field. A strenuous competition is going on between the different species, each of which is endeavouring to crowd out its neighbours, so that the characteristic vegetation of the field represents the state of equilibrium which has been attained by the various plants under the prevailing conditions of soil and climate. The physical texture of the soil has much to do with the nature of the grasses which will establish themselves under the stress of competition. On deep, kindly alluvial pastures, rye grass becomes prominent; on the thin chalky soils of the Downs, sheep's fescue thrives best; on heavy clays where aeration is deficient the creeping rooted bent grass will cover the surface, while sandy droughty soils often become covered with tufts of cocksfoot or brome grass. Manuring, by altering the food conditions in the soil, can effect a great change in the character of the herbage of a given field, and the direction which these changes will take must be kept in mind in any discussion of the application of fertilizers to grassland, since in Great Britain we are seldom dealing with a crop of a pure unmixed grass, like the crops of timothy or blue-grass in America.

An example of the effect of long-continued manuring on the composition of the herbage is afforded by the Rothamsted experiments, where certain plots of old grassland receive the same treatment every year and are mown for hay. Table 61 shows the average yield for fifty-three years, and also the

TABLE 61.—YIELD AND COMPOSITION OF HAY AT ROTHAMSTED.

Plot.	Manure.	Yield of Hay.	Botanical Composition per cent.		
			Gramineæ.	Leguminosæ.	Other Orders.
		Cwt.			
3	Unmanured	21·5	34·3	7·5	58·2
1	Nitrogen only as Ammonium Salts	34·7	77·6	1·4	21·0
17	Nitrogen only as Nitrate of Soda .	35·5	43·8	3·4	52·9
7	Mineral Manures, no Nitrogen .	40·9	20·3	55·3	24·4
4–2	Phosphoric Acid and Nitrogen, no Potash	35·8	91·5	...	8·5
9	Complete Manure, Nitrogen as Ammonium Salts . . .	54·8	91·2	1·3	7·5
14	Complete Manure, Nitrogen as Nitrate of Soda . . .	60·8	88·8	3·7	7·5
11–1	Complete Manure, excess of Nitrogen	66·8	99·2	...	0·8

character of the resulting herbage, as shown by its separation into grasses, clovers, and weeds in 1902, the forty-seventh year of the experiment.

From this table certain facts become apparent. If grass is constantly mown without any return in manure, the resulting impoverishment is shown not only in the small yield but in the preponderance of weeds in the herbage. One-sided manures, which contain only nitrogen or only phosphate, however successful at first, eventually result in increased impoverishment of the land. Nitrogenous fertilizers promote the growth of the grasses at the expense of the clovers. Mineral manures, and particularly potassium, promote the growth of leguminous plants and enable them to make headway against the grasses.

Another consequence follows from these experiments; since any special combination of fertilizers or any method of treatment encourages particular species, the best results in any given field will always be attained by persisting in the treatment selected. For instance, when a field is laid up for hay certain strong-growing grasses get an advantage; when the field is grazed, other grasses of a dwarfer-spreading habit are more suited by the conditions. It is therefore desirable to keep one field for hay every year and another for grazing, rather than alternately to graze and hay the same field, in which case particular grasses are first of all encouraged and then repressed.

Again, we may conclude that manure will be wasted upon a field unless there is a proper herbage to take advantage of it; in dealing with poor grassland it is uneconomical to spend much on manure until by degrees the character of the vegetation has been reformed. With these general principles in mind, we may proceed to the consideration of a few typical cases, which, however, cannot be made to cover all the variations of soil and management to be met with in practice.

Land laid up for hay every year must receive a regular manuring, unless it happens to be rich river meadow or alluvial flat which derives its fertility from the percolating water or the mud deposited during flood time. But if it is ordinary medium grassland, about 1 cwt. per acre of muriate and 2 cwt. of superphosphate should be applied in the early spring, followed by 1 to 1½ cwt. of sodium nitrate as soon as the grass begins to move. On heavy soils, especially on old grassland, basic slag may be advantageously substituted for the superphosphate. At intervals of five years or so the mixture of fertilizers should be replaced by a winter dressing of about 15 tons of farmyard manure. Every five or six years a dressing of lime or limestone should be given. Land that has been but recently laid down to grass should be dunged more frequently. If much cake and corn is fed on the aftermath, the nitrate can be reduced or even omitted.

Seeds hay should not require any manuring; if the land has been properly treated before the seeds were sown there should be enough residue from previous manuring to grow a good crop of mixed seeds. Any active nitrogenous manure will stimulate the rye grass, at the expense of the more valuable clovers, but an application of 1 cwt. nitrogenous fertilizer 2 or 3 weeks before cutting effects a striking increase in the protein content of the hay. A fertilizer is sometimes used in the spring when the land has lost plant severely through the winter, but this is generally a wasteful proceeding, because fertilizers should only be used when there is a crop or the prospect of a crop to utilize them.

Pasture that is of any value to begin with will rarely require general manuring if much cake and corn are fed to the stock fattening upon it in the summer. Lime and phosphates are, however, often deficient on old pastures, and for lack of these constituents the great residues of manure left on the land every year are not adequately realized. For this reason occasional dressings of ground limestone (2 tons per acre) and of basic slag (5 cwt. per acre) are of great value on these lands where cake and corn are fed. The result of the application may not be visible in an increased growth of grass, but the cattle will be found to prefer the manured portions of the field and to thrive better there. This is because of the greater uptake of calcium and phosphorus improving the food value of the herbage. But grazing management is as important as manuring.

Poor pasture in Great Britain may be divided into three classes: (1) poor clay land covered with creeping-rooted bent grass; (2) thin sandy soils covered with sheep's fescue, fiorin, sweet vernal, and soft brome grasses; (3) thin soils near the chalk with an extremely variegated herbage.

As regards the first class of land, the experiments initiated by Somerville at Cockle Park, and extended later to many other clay soils all over the country, show that dung and other nitrogenous manures are worse than useless on such soils. The sound way of improvement is to give them a dressing of 10 cwt. or so per acre of basic slag, whereupon the white clover becomes prominent in the herbage. The grazing is at once improved, and as the nitrogen consumed mostly comes back to the grass, a permanent improvement sets in. Should white clover not appear the season after the basic slag has been applied, a pound of wild white clover seed should be sown and harrowed in. Severe mechanical treatment of the turf is in any case of great assistance in speeding up recovery. After this first dressing of basic slag, the land will steadily improve for five or six years, and a fresh application of fertilizer is called for. By this time the soil will have gained nitrogen through the growth of the white clover, but it will not be wise to trust to basic slag alone for the second dressing. The second and later dressings should be accompanied by about 1 cwt. per acre of muriate to keep the clover vigorous ; and if the land is ever laid up for hay, it will be necessary to use nitrogenous fertilizers pretty freely. As long as a pasture containing a good proportion of white clover is only grazed, it is probable that the nitrogen content of the land will not fall off, but we cannot trust to white clover to make good the large removal of nitrogen in a hay crop.

The thin sandy soils are more difficult to improve than the clays; basic slag exerts but little effect because the soil is too dry to allow it to act very freely, and the soil may be short of potassium. Bone meal has often been recommended for these soils but is too slow in its action to be profitable, and a fertilizer with a good proportion of easily soluble phosphate is better. About 2 to 3 cwt. of such a phosphate and an equal amount of potash salt is the best type of mixture to improve the herbage on these very light soils, but even then the change will be slow and never so pronounced as on clay land, because the tufted deep-rooting grasses which prevail are better able to resist the competition of the leguminous plants. Nitrogenous manures, and particularly dung, may be harmful, encouraging the coarse herbage.

On the thin chalky soils, nitrogenous manures are valuable, and a pasture may be so improved as to carry more stock in the current season by the application of 3 or 4 cwt. per acre of a mixed fertilizer, containing 3 parts of superphosphate, and 1 each of muriate and ammonium sulphate. But for the creation of a good pasture on the thin chalk soils, dung is the most essential manure; as much farmyard manure as possible should be spared for the grassland and a hay crop taken the season after the application; then it should be grazed and, if necessary, helped during the grazing by the mixture specified above.

But it must always be remembered that on the thin dry soils, whether chalk or sand, only a very limited expenditure on fertilizers is likely to be repaid; large applications of manure will be certainly wasted, but it is possible gradually to build up better pastures by repeated small applications of the nature described.

The treatment of pasture thus indicated—the application of basic slag or other phosphatic fertilizer which by its encouragement of white clover gradually builds up a more fertile soil growing an increased amount of grass as well as clover—is a slow regenerative process which does not call for much outlay of capital. Though it effects a manifest improvement in the pasture it cannot produce really rich grazing or anything in the nature of a "fatting" pasture.

However, the improvement of an old matted pasture by the addition of fertilizers and lime and light harrowing, takes a long time. Severe mechanical treatment of the turf is needed to open it up and allow water and manures and roots to penetrate down to the mineral soil. Many farmers have been loath to use the heavy implements designed for this purpose and without which renovation is limited and extremely slow. Much experimental work all over the country has demonstrated that by far the most effective method is to plough and, with the aid of subsidies for ploughing up old grass and for lime, and also temporarily for fertilizer, striking results have been obtained in the conversion of much worn-out old pasture into good grazing.

The conditions are variable, of course, but on the upland areas of the east of Scotland the procedure has been to plough in such a way that the turf is completely inverted and the furrow seams closed as far as possible to destroy the chances of survival of the old vegetation. Enough lime is then spread and disced in to reduce acidity, and the soil is rolled to obtain a firm seed-bed; this is followed by a dressing of slag to correct phosphate deficiency and a light dressing of nitrogen to give the young grass a good start; a suitable seeds mixture is sown and, provided the soil has not been allowed to dry out, the land may be ready for grazing in about 8 weeks. Experiments with sheep have shown that the carrying capacity is raised to such an extent that live-weight returns may be increased as much as eight-fold; sometimes indeed a section of the new grass could be set apart for haying.

This method of ploughing and direct reseeding has much to commend it although some farmers still prefer to start with a cereal nurse crop for the seeds, or with rape. The problem is bound up with such things as the nature of the site, mode and time of cultivation, amelioration of different soils and selection of suitable seed mixtures, all of which have engaged the attention of investigators. It is a matter of immense economic importance and these remarks are intended simply to draw attention to the possibility of increasing the productivity of areas of worn-out pasture.

The improvement of grassland has been approached from another point of view, beginning with second-class grassland rather than with the upland grazings to which slagging is most applicable. The theory was first worked out by Professor Warmbold of Hohenheim as a means of providing for a dairy herd during the 1914–18 war, when supplementary feeding stuffs containing protein were not obtainable but synthetic fertilizers were being made in quantity. The essence of the method consists in the repeated applications of nitrogenous fertilizer and the regulated grazing of the land so that the grass is never allowed to grow beyond a height of six inches or so. As is well known, any good pasture exhibits its highest feeding value in the months of May and June when the grass is making its first growth and is richest in protein and ash constituents. Later on the capacity of the pasture to carry stock falls away and though abundant growth is often renewed in September the grass has not such a high feeding value. Warmbold's system aims at renewing the rich growth of high feeding value by applications of active nitrogenous fertilizer. The method consists in dividing the available grazing into a series of 5 to 8 paddocks, each 4 to 5 acres in extent. At the outset in January the whole receives a basic manuring of 5 cwt. of superphosphate and 5 cwt. of potash salts per acre with 10 to 20 cwt. of ground lime or calcium carbonate if necessary. As soon as the weather is open in February the land receives 1 cwt. per acre of ammonium sulphate, this application being renewed a month later.

The growth is generally so rapid under this stimulation that the milch cows or cattle can be brought out a month earlier than on the unmanured land, say in early April. The stock are put on the first plot only, which is heavily stocked up to two and even three cows per acre. The best of the grass will be eaten off in three or four days and the stock must never be allowed to remain on a stale pasture but are shifted on to the next plot as soon as they begin to wander in search of food. The grazing of the first plot is then completed by turning on dry cows or rough bullocks, or sheep and horses; in another few days it should be closely eaten down, by which time the two lots of stock are ready to move forward one more plot. The cows in milk always go forward to a fresh plot and are followed up by the dry cows. Meantime the first plot, as soon as it is cleared, is well harrowed to distribute the dung and is given a second dressing of 1 cwt. of ammonium sulphate per acre. Growth is immediately renewed and the plot is getting ready for grazing again when its turn comes round. The rotational grazing then proceeds, the cows in milk being moved as soon as they have taken the best off the fresh successive plot. How long they will remain will depend on the weather and the current rate of growth of the grass. In very favourable seasons the growth may get ahead of the stock; in such cases one or more of the plots may be cut for silage as soon as it has reached the allotted limit of growth, or more stock may be introduced. The leafy grass silage makes very valuable fodder for winter use, or as a supplementary ration in case of drought. A short drought is met by diminishing the length of stay on each plot; a long-continued spell in which the growth comes to a standstill can only be met by artificial feeding or by turning the stock over to some reserve grazing until the plots recover. But the highly manured plots last out better in a drought than does unmanured land; again profitable growth begins earlier in the season and lasts later.

Experience shows that in this country ordinary second-class grassland treated in this way rapidly takes on the aspect of a good pasture and is capable of supplying the maintenance ration of a dairy cow together with the material required for the first and even the second gallon of milk per day. The drawbacks to the method are the initial cost of fencing into small paddocks and supplying water, the greater skill and care required in moving on the cows the moment the grazing is past its best, and the difficulty of securing an even distribution of the large amount of droppings. Some consideration is also required to adjust the system of the stocking of the farm and to balance the intensive grazing with the right proportion of reserve grazing and hay land, so that the farmer is never caught heavily overstocked in a period of drought. The method answers best on a large farm where the occupier has a margin of stock to deal with any excess of grass and a margin of land to provide for the stock in times of long drought. Sheep have been grazed on the same plan but the nitrogenous manuring cannot be repeated so often and the sheep as a rule thrive better with a more extensive run.

The principles of the system have been corroborated by the examination of the grass cut at regular short intervals after the high manuring. If never allowed to grow beyond grazing height it is found to maintain throughout the season the high content of nitrogen and ash and the high digestibility that characterizes the May and June growth. In fact it has been shown that if such short grass is cut and dried artificially it is equivalent in feeding value to cake.

When land has been newly laid down to grass, there often comes a very critical period from its fourth to its seventh year, especially on stiff soils and when the first two or three crops of grass have been fed off by store stock only. At that period the leguminous plants have begun to die away, and the grasses have lost vigour because the plant food that had been rendered available by the tillage has become exhausted. The mechanical condition of the soil has also deteriorated because as yet little humus has been accumulated. Applications of basic slag have less effect than usual on such young grassland; there are no residues of past growth to be set in action by the lime of the basic slag. What is wanted is either farmyard manure or applications of a complete fertilizer such as has been described above. Better still, the land should be carefully pastured, the sheep should not be allowed to eat too closely, and should be fed with cake or corn to enrich the land.

Hops

Few crops are so liberally manured as hops; potato land may perhaps receive as much in any one year, but on hops the expenditure for fertilizers is high year after year. The hop plant shows no special requirements, so that it is the needs of the soil rather than the crop which should determine variations in the character of the manure. The manurial treatment of hops should begin with a liberal use of dung, and most hop growers either buy it in quantities from London or other large towns, or fatten cattle or pigs in order to make enough for their requirements. As much as 40 tons per acre are sometimes employed and that year after year, but one such application every third year will be sufficient to maintain the requisite soil texture, and in the intervening years the necessary plant food can be more cheaply obtained in other forms. The subsidiary manures for hops are of the most varied nature, but shoddy in some form or other is a highly favoured substance, and should be applied at the rate of from 1 to 2 tons per acre, according to its richness in nitrogen, in the autumns when dung is not being used. When the ground is first worked in the spring the more active fertilizers should be applied, 6 cwt. per acre of fish or meat guano or of rape dust, with about 4 cwt. per acre of superphosphate, or 3 cwt. of steamed bone flour or phosphatic guano, will then carry the crop through. Many growers are in the habit of using a further dressing of rich guano or active nitrogenous manure when the hops are coming into burr, but this is probably unwise, as it induces late sappy growth, very susceptible to attacks of blight. A good coat of dung at this time is, however, of great value, especially on young hops, but its immediate action is more as a mulch than a fertilizer. Potassic manures are required only on the light sandy or chalky lands; in such cases they should be applied in the winter or early spring. Phosphates are, however, essential; on strong soils, as much as 10 cwt. per acre of basic slag may be applied in the winter in place of the superphosphate specified above.

Fruit Plantations

Many special problems arise in deciding how best to manure fruit, for there is so much variation in climate, soil type, system of management, and nutrient requirements of different fruits; there is also the importance of maintaining satisfactory quality when yields are increased. The grower must learn from experience to diagnose nutritional disorders and to adjust the manuring to correct them. The problem of manuring is different under the arable system, which has been more common in the drier eastern districts, from that under the grass system, which has been the practice in the moister western parts of the country. The former may consist in taking market garden crops between the rows of fruit, which means that nitrogenous manuring is practised on too heavy a scale to obtain the best results from the trees. On lighter soils particularly, a potassium deficiency becomes apparent and difficult to correct. It is necessary therefore to plant those types which can best stand up to the soil conditions, and plums and soft fruits are in general best suited to benefit from the good moisture conditions and plentiful supplies of nitrogen. But experimental work by the Long Ashton and East Malling Research Stations has demonstrated that potassium is very often a limiting factor in healthy growth; red currants and gooseberries seem to be particularly sensitive in this connection and market garden conditions are not suitable for the best quality dessert apples.

In contrast to this, grass culture may be responsible for a shortage of available nitrogen. To remedy the condition, such heavy dressings of nitrogenous fertilizers may have to be used as to make the grass unfit for grazing; at the same time tree growth is not so good when the grass is removed as hay. The solution lies in careful management of the stock and this usually ensures that the land is adequately manured for the fruit. Poultry managed on the fold or movable pen system, has been a successful method of dealing with the difficulty.

Farmyard manure has generally proved to be a most efficient manure for fruit; where an adequate and continuous supply of water is required during the fruiting season, it is only to be expected that a bulky organic manure would be particularly valuable. There is no evidence on the other hand that the other organic manures like hoof-and-horn meal, meat meal and guano are better than inorganic nitrogenous and phosphatic fertilizers. With respect to potassium fertilizers, chlorides may cause injury and, generally speaking, the sulphate is to be preferred. Potassium deficiency is fairly common in the chief fruit areas of this country and it is not readily corrected. Phosphate deficiency on the other hand is a rare occurrence.

Tropical and Sub-Tropical Crops

It is very difficult to lay down any general rules for the manuring of tropical and sub-tropical crops, because the conditions of soil and climate are subject to such extreme variations that entirely different methods of treatment have to be pursued in different countries. Certain general principles may, however, be indicated, to be taken into account whenever any scheme of manuring has to be tentatively adopted in practice. All the processes by which the insoluble constituents of plant food in the soil are rendered available for the plant are greatly accelerated in tropical soils, always provided they contain a sufficiency of water. The decay of organic matter takes place with extreme rapidity, so that the humus content of cultivated soils is quickly reduced unless means are found of repairing the losses; for the same reason all organic manures containing nitrogen are both more quickly and more completely utilized by the plant than they are in temperate soils. The higher temperature of the soil water and the greater production of carbon dioxide in the soil, also result in a more rapid weathering of the mineral constituents of the soil, so that the reserves of plant food are either more available or are seriously depleted by leaching.

The incidence of rainfall must be closely studied. No manure can be effective when the soil is either dry or waterlogged; and as the nitrogenous manures cannot be expected to persist very long in the soil, their application should be timed so as to be followed by a period of growth with neither excessive rain nor a dry soil.

Sugar Cane.—A large number of experiments have been conducted with sugar cane, and though the results naturally vary in the different countries, certain general conclusions can be drawn. Before planting, a comparatively slow-acting nitrogenous fertilizer should be used, either the equivalent of farmyard manure or some seed residue like castor pomace. For the rattoon growths more active forms of nitrogen are desirable, but excess of nitrogen must be avoided, as it induces late cane and an impure juice. Recent studies in the Deccan Canal Tract have led to the recommendation of green manuring plus three top dressings of nitrogen partly as ammonium sulphate and partly as oil cake. The amount of nitrogen which can profitably be used depends upon the variety of cane grown but, for January plantings of the modern high yielding varieties, about 300 lb. N per acre is satisfactory; about 15 lb. of this total should be readily available nitrogen at germination and 50 to 100 lb. is best for tillering. On many soils applications of potassium salts are very effective. Phosphates are less needed, though superphosphate is often valuable on black alluvial soils.

Cotton.—Cotton responds freely to fertilizers, and there is evidence that the fertilizer should be a mixed one but mainly phosphatic. About 4 cwt. per acre of superphosphate and 2 cwt. of cotton seed meal or some equivalent organic source of nitrogen, should be ploughed in before sowing, and this may be followed up by ½ cwt. of a more active nitrogenous fertilizer like ammonium sulphate or sodium nitrate when the crop has begun to grow. Potassic manures are only required on certain soils of a light type. Sodium is of great value in increasing the yield of seed cotton.

Tobacco.—Tobacco is a crop requiring comparatively rich land, and the fertilizers should supply chiefly nitrogen and potassium, phosphates being less required. Too great an amount of nitrogenous fertilizer should not be used, or the quality of the leaf will fall—up to 50 lb. per acre is safe; and ammoniacal manures should be avoided, as they result in the leaf burning badly. Before planting out the tobacco, 200 to 300 lb. of an organic nitrogen compound—cotton seed meal or castor pomace—200 lb. of superphosphate and 100 lb. of potassium sulphate should be applied, followed by 100 lb. of sodium nitrate when the plant is growing. Potassium appears to be essential, and may be given as nitrate, carbonate or sulphate.

Tea.—Being perennial, the tea plant requires neither heavy nor active manuring; it is also very important to maintain both the proper habit of growth of the plant and the quality of the leaf. If any large amount of nitrogen is employed an excessive development of weak vegetative shoots takes place on the bush, and the plant suffers in ensuing seasons. The fertility of a tea garden as regards nitrogen can be maintained by carefully burying the lighter prunings and weeds, and by digging in from time to time leguminous plants which have been grown between the rows, cut down, and allowed to wither and rot somewhat. By adding basic slag at the rate of about 2 cwt. per acre the residues thus utilized are balanced by phosphate, and the lime of the basic slag is beneficial in keeping the soil healthy and in assisting the decay of the organic matter, without reducing the acidity too much. Tea grows best under acid conditions.

The results of a study of the responses of tea to manurial treatment, over a period of four 3-year pruning cycles in Ceylon, have recently been reported by Eden. His chief conclusions were that although potassium affects the composition of the leaf, it does not influence the yield of dry matter produced. The response in dry matter to phosphate equivalent to 30 lb. P_2O_5 was similar to that for crops grown in temperate regions, but larger dressings were no more effective. The effect of nitrogen was most striking and reached a maximum of 6 lb. dry matter per lb. N applied. This is low compared with the responses by cereal and root crops here, but is important considering that the tea was grown on leached upland soils and that the efficiency of the nitrogen increased as the bush grew larger in each cycle and also with successive cycles. This is presumably because the absorption of nitrogen is rapid and that, after pruning, either the activity of the roots is low or the plant is unable to deal with nitrogen.

Gardens

In an ordinary way gardens require little fertilizer since they receive a superabundance of stable manure until the soil often becomes over-rich in nitrogenous residues. Under such conditions the only fertilizer wanted will generally be some form of phosphatic manure, but potassium is also required on occasion to secure a properly balanced growth in the crops. Superphosphate may be used on the loams, basic slag on the strong soils, steamed bone flour or phosphatic guano when the soil is sand or gravel, and about ¼ lb. per square yard of one of these fertilizers should be dug in with the farmyard manure on those portions of the ground which come to be dunged in the usual rotation. Sodium nitrate or nitrochalk is often valuable to push on early lettuce, cabbage, peas, etc., in a backward spring; it may also be applied with advantage to asparagus and celery and various other crops at suitable stages of growth. The many compound garden manures sold under fancy names may be good fertilizers but their cost is often excessive, even considering the small parcels in which they are made up. Where a mixed fertilizer is required an ordinary potato manure is usually quite suitable. If stable manure is unobtainable the humus of the soil should be maintained by digging in as much organic matter as possible—weeds, grass clippings and vegetable refuse—preferably after composting, or by growing mustard on any land that is not wanted for a short time, and digging the green crop in.

It should not be forgotten that lawns which are constantly cut must become greatly impoverished if the cuttings are removed. Ammonium sulphate is valuable as it encourages the growth of the finer grasses at the expense of clover and the broad-leaved weeds. Frequent applications at the rate of 1 to 2 lb. per 100 sq. yd. should be made during the summer months. In course of time the soil will become acid and it will be necessary to replenish its supply of calcium, phosphorus and potassium without, however, providing conditions suitable for undesirable species.

For seed boxes and potting purposes it is usual to prepare a mixture of rich loamy soil, peat or leaf litter and coarse sand in order to obtain a medium that will hold water but drain freely and encourage root development. Some lime and superphosphate are desirable to ensure that conditions are not too acid and that phosphate is available; other fertilizers such as hoof-and-horn meal and potassium sulphate are incorporated or applied later according to the grower's requirements. The recommendations of the John Innes Institute have found great favour.

CHAPTER XIII

THE VALUATION AND PURCHASE OF FERTILIZERS

Unit values—Fertilizer prices (1955)—Regulations—Mixtures—Manurial residues—Compensation values of feeding stuffs.

IN buying fertilizers the farmer will generally have a considerable choice between materials of different origin and composition which will so far serve the same purpose that their relative price becomes the most important factor in determining the purchase of one or the other. For example, should an active nitrogenous manure be needed it is often a matter of indifference whether sodium nitrate or ammonium sulphate is used; among phosphatic manures the choice may be between superphosphate and basic slag; or, to take a case where even fewer secondary considerations enter, practically nothing but relative cheapness need determine a decision between such materials as fish and meat guanos or rape cake. But since all these materials possess different compositions, a method of valuation must be found which will reduce to a common basis the cost of the actual fertilizing ingredients alone.

Unit Values

In this country, dealers in fertilizers are obliged to give the analysis of their products in terms of nitrogen (N), phosphoric acid (P_2O_5) and potash (K_2O). Some merchants still give figures for ammonia in addition to N, and for tri-calcium phosphate in addition to P_2O_5, in order to adhere to an older convention and assure the purchaser that the lower figures under the latest regulations do not mean that the fertilizer is inferior in quality. It is simply a question of 14 parts N being equivalent to 17 parts of ammonia (NH_3) and 142 parts P_2O_5 being equivalent to 310 parts tri-calcium phosphate.

The present system is archaic and causes endless confusion. As explained in Chapter V, P_2O_5 is not phosphoric acid but phosphoric anhydride or phosphorus pentoxide; it would simplify matters to express all phosphatic fertilizers in terms of phosphorus (P). Similarly, to describe muriate of potash —the old-fashioned name for potassium chloride—as containing 60 per cent. potash, means that the essential element potassium is present in the potassium chloride in such a quantity that, if it were combined with oxygen, the potash or potassium oxide (K_2O) formed would amount to 60 per cent. of the fertilizer. It is almost incredible that such a complicated manner of stating the fertilizer content of potassium (K) should have survived so long. Unfortunately, these conventions are so firmly established in the fertilizer trade all over the world that the prospects of reducing what might be a comparatively simple matter to simple terms are not bright.

Meantime then, the cost of fertilizers is expressed in terms of units of N, P_2O_5 and K_2O; the cost of a unit, consisting of 1 per cent. of the constituent in a ton, is found by dividing the price per ton of the fertilizer by the percentage of the constituent in question.

The prices of fertilizers are subject to variation from month to month; the figures given in Table 62 are from market quotations early in 1955; but the purchaser may recover subsidy for nitrogenous and phosphatic fertilizers. The cost of fertilizers is also included in grants under the Hill Farming (1946) and Livestock Rearing (1951) Acts.

When two constituents are present the unit values have to be obtained by a little adjustment and do not possess quite the arithmetical certainty which characterizes the single constituent fertilizers. Supposing it is desired to find the

TABLE 62.—FERTILIZER PRICES (1955).

Name of Fertilizer.	Guarantee per cent.			Price per ton. £ s. d.			Price per unit. £ s. d.		
	Nitrogen N.								
Sulphate of Ammonia .	20·6			17	17	6		17	4
Nitrate of Soda . .	16			26	5	0	1	12	9
Nitrochalk . . .	15·5			15	14	0	1	0	3
Dried Blood . .	12			55	0	0	4	12	0
Hoof-and-horn . .	13			45	0	0	3	9	0
	Phosphate P_2O_5.								
Superphosphate . .	18			13	4	6		14	8
Basic slag (high soluble)	16			7	5	0		9	1
,, ,, ...	13			6	8	6		9	11
,, ,, ...	10			5	13	6		11	4
Mineral phosphate .	29			11	0	0		7	7
Triple superphosphate .	47			35	19	0		15	4
	Potassium K_2O.								
Muriate of potash . .	60			19	4	0		6	5
Sulphate of potash .	48			22	0	6		9	2
	Compound.						Calculated value.		
	N.	P_2O_5	K_2O						
Potassic phosphate .	...	20	10	14	8	0	10	16	0
Bone meal . . .	3·8	20·5	...	27	15	0	14	1	0
Steamed bone flour .	1	30	...	28	10	0	16	0	0
Chilean potash nitrate .	15	...	10	29	0	0	27	16	0
Concentrated compound	12	12	15	25	14	0	24	0	0

value of the phosphate unit in those fertilizers which also contain a little nitrogen, it is necessary to make a deduction from the price of the fertilizer for the nitrogen present, assuming the latter to have approximately the value calculated from a purely nitrogenous fertilizer. For example, if a steamed bone flour containing 1 per cent. N and 30 per cent. P_2O_5 is quoted at £28 10s. per ton ; the nitrogen at 20s. per unit would be worth 20s., which deducted from £28 10s. leaves £27 10s. for the phosphate. Dividing this figure by 30 (the percentage of P_2O_5), we get 18s. 4d. as the price of the unit of phosphoric acid, which is twice the cost of the unit of

P_2O_5 in a high grade of basic slag. Another way of assessing the value of a compound fertilizer is to calculate how much it is worth in terms of the unit values of the constituents in a "straight" fertilizer. For example, the potassic phosphate listed in Table 62 contains 10 per cent. K_2O and 20 per cent. P_2O_5 which are worth 10 × 6s. 5d. and 20 × 7s. 7d. at the unit values of K_2O in muriate of potash and P_2O_5 in mineral phosphate. The total is £10 16s. to the nearest shilling as compared with the price of £14 8s. per ton. In the case of potash nitrate it is reasonable to calculate in terms of the unit value of N in nitrate of soda ; so that potash nitrate is worth 15 × 32s. 9d. + 10 × 6s. 5d. or £27 16s., using the unit value of K_2O in muriate of potash. With phosphatic fertilizers it is not possible to be certain which unit value is the best to use in making such a calculation. The value of P_2O_5 in superphosphate is obviously most suitable for water-soluble phosphates but for manures like guano or bone products it may be better to use the slag or mineral phosphate values. In Table 62 the calculated values for bone meal and steamed bone flour are based upon 20s. per unit N and 10s. per unit P_2O_5. It is obvious that the cost of the plant food in both products is excessive by these standards. If the value of the P_2O_5 in the various waste organic manures is fixed at a figure comparable to that for slag or superphosphate, then it will be found that the price of nitrogen is very variable and sometimes reaches fantastic levels. In this particular case of bone meal it is actually 90s. per unit ; the 20·5 units P_2O_5 cost 205s. at 10s. per unit which leaves 350s. for the 3·8 units N. The N unit at 92s. in dried blood and 69s. in hoof-and-horn meal are other examples of the values placed upon organic forms of nitrogen.

Such high prices do not represent the intrinsic values of these manures—their relative effectiveness in producing crops —but are due rather to market conditions in which the supply cannot meet the demand. Rape dust, for example, was at one time almost as cheap a source of nitrogen as nitrate of soda and established itself in the favour of hop growers and others who continued to purchase it in the face of a falling supply and an increase in price. The special merits of the bulky organic manures in preserving or improving soil texture have already been discussed and the farmer may feel impelled to pay heavily for them when he cannot obtain enough farmyard manure; but it is difficult to justify the need to pay high prices for the more concentrated organic manures, which are not used in quantities large enough to affect soil texture, simply because they provide a slow but steady supply of plant nutrients over a long period. An increase in the practice of using relatively small dressings of readily available plant food at suitable stages in the development of the plant would appear to be a much cheaper method of getting the same results.

While it is impossible to reduce all fertilizers to a common basis on account of differences in their mechanical condition and speed of action, the unit system of valuation is of great assistance in comparing materials of similar origin or behaviour. It often reveals discrepancies in prices and is of great importance to the farmer in coming to a decision about a purchase.

Regulations

The trade in fertilizers is regulated by the Fertilisers and Feeding Stuffs Act of 1926, according to which "It shall be the duty of every person who sells for use as a fertilizer of the soil . . . any article included in the first column of the First or Second Schedule to this Act, whatever may be the name under which the article is sold, to give the purchaser on or before delivery, or as soon as reasonably practicable thereafter, a statement in writing . . . containing the following particulars:

(*a*) the name under which the article is sold,

(*b*) such particulars (if any) of the nature, substance, or quality of the article as are in relation to the article mentioned in the second column of such schedule."

This obligation does not apply "to sales of two or more articles which are mixed at the request of the purchaser before delivery to him" or to sales of quantities of 56 lb. or less under certain conditions. The sale of an article under a name specified in the Fourth Schedule acts as warranty that the article accords with the appropriate definition in the schedule. The Schedules set out for each fertilizer the particulars that must be given, e.g. for rock phosphate the amount of phosphoric acid and the amount that will pass through the prescribed sieve; for guano the amounts of nitrogen, phosphoric acid and potash. Certain limits of variation are laid down, e.g. if the statement concerning nitrate of soda gives its content of nitrogen at 15·5 per cent. the warranty is not infringed if analysis only shows 15·2 per cent. of nitrogen for a variation of 0·3 per cent. is allowable.

Under the Act of 1926, the duty of administration is placed upon the local authority, county council or county borough, each of whom must appoint an official analyst and an official sampler. On notice being given and a required fee paid the official sampler will attend and take samples of a consignment of fertilizer, thus relieving the farmer of any trouble and any disputes as to proper sampling or packing and consignment of the samples when taken. The sample must be taken within 14 days of delivery of the article or of the receipt of the warranty. The provisions of the Act give the official sampler power to enter the premises of makers and sellers of fertilizers and there take samples. Thus the trade is to some extent being inspected and controlled by an administrative service not set in motion by the purchaser. When the analysis reveals a deficiency not covered by the limits of variation, the matter may be settled between purchaser and seller by an adjustment of price, but in certain cases of false description or the inclusion of harmful materials in feeding stuffs the county council may, after obtaining the consent of the Ministry of Agriculture, or appropriate body in Scotland or Northern Ireland, take proceedings against the seller.

Now that every farmer in the country can so readily and cheaply obtain an analysis of any fertilizer he purchases, for besides the county analyst some of the large agricultural societies have retained an analyst for the assistance of their members and the advisory services also undertake analyses for the farmers resident within the area they serve, he ought to get analyses made of every purchase of certain classes of fertilizer, if he has any regard to the economical conduct of his business. With very few exceptions the manufacturers and vendors of fertilizers supply the farmer with the material they have undertaken to sell. As regards the pure unmixed fertilizers, standard articles made on an enormous scale, such as nitrate of soda, sulphate of ammonia, superphosphate, chloride and sulphate of potash, any farmer dealing with a reputable firm may count on getting what he pays for, because these materials do not vary in composition unless they have been deliberately falsified after they have left the wholesale hands. But with waste materials so many different samples exist of varying composition, and so easily may even a single cargo show differences in passing from one part to another, that the farmer will always be wise to check his purchases by an analysis, not of course of the sample that may be submitted to him before purchase, but of the consignment on arrival.

The farmer should buy his fertilizer on the strength of the guarantee from the vendor, before he gives his order and, of course, he should work out a valuation by the method described earlier in the chapter; he should then be careful to see that the invoice agrees with the guarantee on which he bought. In case of doubt, he should have an analysis made of a sample drawn from the bulk delivered.

Mixtures

In order to exercise to the full his power of buying in the cheapest market prevailing, it is clearly necessary for the farmer to know the kind of fertilizers he wants for the crop in question, so that he can compound the available materials in the right proportions if he is not entirely satisfied with the necessarily limited range of fertilizers already mixed by the manufacturer. For example, every merchant's catalogue will show examples of turnip manures, barley manures, mangold or grass manures, containing such mixtures of nitrogen, phosphoric acid, and potash as experience has shown to be generally suitable to the crops in question. In such cases the farmer gets the advantage of the knowledge of the merchant and also obtains a carefully mixed fertilizer of even composition throughout, which can be distributed without further trouble. Such mixtures, however, can only represent a certain average adaptability to the crop and may not take into account either the particular kind of land or the condition it has been left in by previous cropping.

The farmer who has really made himself acquainted with the theory of manuring and with the special conditions of his own land can often manure both more cheaply and more effectively by purchasing unmixed fertilizers. These he must either sow separately, or by paying a little extra to the merchant he may get made up whatever mixture he desires before delivery. The average farmer has not the facilities for making as good a mixture as the merchant but, if it is necessary, it should not be a matter of great difficulty to make up a sufficiently accurate compound on the farm. A suitable weighing machine is required and a cemented or paved floor on which lumps can be crushed. The heaps of separate manures should be weighed out and thrown into a common heap by alternate shovelfuls; the mixture should be then passed through a half-inch screen and the lumps broken down with a wooden rammer or the back of a shovel, the resulting heap being cast down and remade two or three times until it is uniform in appearance.

It must be remembered that certain fertilizers cannot be mixed together without setting up reactions which are either wasteful or render the mixture difficult to work. Basic slag or calcium cyanamide should not be mixed with sulphate of ammonia or guano or any other fertilizer containing ammonium salts, because the active lime sets free ammonia, which escapes as a gas. Superphosphates cannot long remain mixed with nitrate of soda without setting free a certain amount of nitric acid, which is both wasteful and injurious to anyone handling the mixture. It is, however, safe enough to make up the mixture and sow it straight away; the nitric acid only begins to be in evidence when the mixture is left in a heap or in bags overnight, or when it is sown from a machine which has some moving part working in the manure. Most mixtures containing superphosphate will turn into a paste round machine parts working in the material. Kainit and superphosphate will also begin to set free hydrochloric acid if they are left long together.

Superphosphate, either mineral or bone, can be safely mixed with ammonium sulphate or guano or any of the fish or meat compounds, but compounds containing calcium are liable to bring about reversion of the water-soluble phosphate. The only nitrogenous fertilizers that are suitable for mixing with basic slag are sodium or calcium nitrate or calcium cyanamide. Of course lime, like basic slag, should never be mixed with ammonium sulphate or any of the guanos or organic nitrogen fertilizers; calcium carbonate may be used if the mixture is applied quickly.

Manurial Residues

One other question of valuation and price comes into play in connection with fertilizers, and that is the value of the residues left behind in the soil after one or more crops have been grown. The provisions of the Agricultural Holdings Act of 1900 awarded the tenant compensation for any unexhausted fertility he had brought to and left behind on the holding. A tenant, for example, who has given his grassland a dressing of 10 cwt. per acre of basic slag and then leaves his farm within the following two years will by no means have reaped the full benefit the land has derived from its treatment. On the other hand, a tenant who has used sodium nitrate to grow his last crop of oats or wheat, and then sells the grain produced, will have obtained all that the manure can return; no nitrogen will be left behind in the soil for the benefit of the succeeding tenant.

It is thus necessary to consider each fertilizer separately

and attach some value to the residue left behind after one or two crops have been grown since its application. It cannot, however, be said that sufficient data exist for the compilation of a rigid scale of compensation; it is not enough to estimate what proportion of the fertilizing materials applied are removed by the crop, and then assume that the remainder is available for future crops. Experiments already quoted will have served to show that residues of slow-acting fertilizers, such as farmyard manures and shoddy, are very far from being wholly recovered even after such long intervals of time as would render any compensation quite out of the question. To a certain extent, again, the value of the residue left by a particular fertilizer will be determined by the nature of the land and the crop to which it has been applied; kainit applied to heavy clay land might not add to the value of the land; on the other hand, the benefit derived from the application of fertilizers to grazing land is largely cumulative, depending upon the change it effects in the botanical composition and quality of the herbage, so that the benefit may be greater at the end of the third or fourth year after application than earlier. Certainly no *a priori* rules for compensation based upon purely theoretical considerations can be laid down; any scale of compensation must be based upon experiments only and must always be considered as approximate and subject to modification according to the particular conditions of soil and cropping.

The crude practice adopted by many valuers of allowing to the outgoing tenant half the cost of the purchased fertilizers he had applied during the last year of his tenancy, could find little or no justification, and in the case of such substances as sodium nitrate and ammonium sulphate was obviously unjust to the incoming tenant, unless the manure had been applied to root crops which had been consumed on the farm.

Various experimental data have been reviewed to compute the residual effects of various materials as set out in Table 63.

In the case of liming materials, it is recommended that one-eighth of the cost on the farm be deducted for each year

TABLE 63.—COMPENSATION FOR FERTILIZERS APPLIED AS FRACTION OF CURRENT UNIT PRICES.

Component.		After growing season.		
		1st	2nd	3rd
N	In organic compounds (excepting dried blood)	1/2	1/4	nil
P_2O_5	(*a*) Soluble in water or citric acid	2/3	1/3	1/6
	(*b*) In bone products	1/2	1/4	1/8
	(*c*) Insoluble forms	1/3	1/6	1/12
K_2O	Total	1/2	1/4	nil

after application. It was common practice at one time to regard the wastage as being equivalent to 4 cwt. of lime or 7 cwt. calcium carbonate per acre per annum. Both methods may be criticized on the ground that the annual loss is not likely to be uniform in successive years or under different climatic and soil conditions; but only an approximation to the truth is possible. No compensation is allowed for the various inorganic forms of nitrogen.

Such a table can be no more than a guide to valuers as to what may reasonably be expected under normal conditions. The valuer must himself take into consideration whether the fertilizer was suited to the soil and the course of cropping, and any other circumstances that would affect the value of the residues to the incoming tenant.

In the somewhat analogous case of the compensation to be paid to the outgoing tenant for the fertility he leaves on the farm from foodstuffs purchased and consumed during the last years of his tenancy, it is possible to draw up a fairly satisfactory scale. The custom in many parts of the country was to allow the outgoing tenant one-half of the cost of the foodstuffs he had brought on to the farm during the last year of his tenancy, but such a system has obviously no scientific basis. The price of a given feeding stuff is determined by its value as food, not as manure; oil or fat, for example, is a costly constituent of a feeding stuff and yet leaves no fertilizing residue behind. Many foods, e.g. maize and rice, consist mainly of carbohydrates and contain an unusually small proportion of nitrogen and ash which would wholly or in part add to the fertility of the farm. The proper basis is to ascertain the nitrogen, phosphoric acid, and potash contained in each class of feeding stuff; an estimate can then be formed of how much of each of these is likely to reach the manure, and a valuation made of these latter quantities at the current rates. The difficulty lies in estimating the proportion in which the fertilizing constituents will reach the manure; for example, it has been shown (Chap. VIII) that of the nitrogen fed to an animal anything up to 15 or 20 per cent. will be retained by the animal, and of the rest that is excreted as much as one-half may be lost in making the dung. Taking a general average from the experiments quoted, it will be seen that about one-half of the nitrogen in the food is likely to find its way to the land in the dung produced under ordinary conditions of farming. If the manure is carelessly managed the losses will be greater; on the other hand, if the food is consumed directly on the land the only loss should be the amount retained by the animal. Similarly, milch cows will retain more than fatting bullocks, young growing stock than work horses; and again, these variations will be set off by the fact that both milch cows and young stock are largely fed on the land.

Taking these and other considerations into account J. A. Voelcker and Hall constructed a scale of compensation in 1902 for purchased foods. The basis was that one-half the nitrogen, three-quarters of the phosphoric acid, and the whole of the potash in the food consumed during the last year of the tenancy is found in the dung, while of the food consumed in the previous year only one-half of these latter values will remain on the farm. Tables were published giving the manurial values per ton of various foodstuffs.

The matter was reconsidered in 1927 by a Scientific Committee and a Joint Committee of chemists and representatives of various professional bodies, and certain

TABLE 64.—THE COMPOSITION, MANURIAL AND COMPENSATION VALUES OF FEEDING STUFFS.
(Revised August 1954 by the Scottish Standing Committee.)

Foods.	Valuation per Ton as Manure.																D Compensation value for each ton of food consumed.			
	A Nitrogen.					B Phosphoric Acid.					C Potash.									
	Per cent in food.	Value at 13s. per unit.		40% of value to manure.		Per cent in food.	Value at 7s. per unit.		Two-thirds of value to manure.		Per cent in food.	Value at 6s. per unit.		Four-fifths of value to manure.		(1) Before one crop has been grown or removed.		(2) After one crop has been grown or removed.	
		s.	d.	s.	d.		s.	d.	s.	d.		s.	d.	s.	d.	s.	d.	s.	d.
Decorticated cotton cake	6·90	89	8	35	11	3·10	21	8	14	5	2·00	12	0	9	7	60	0	30	0
Linseed cake	4·75	61	9	24	8	2·00	14	0	9	4	1·40	8	5	6	8	40	8	20	4
Coco-nut cake	3·40	44	2	17	8	1·40	9	9	6	6	2·00	12	0	9	9	33	11	17	0
Earth-nut cake	7·62	99	0	39	7	2·00	14	0	9	4	1·50	9	0	7	2	56	1	28	0
Beans	4·00	52	0	20	9	1·10	7	8	5	1	1·30	7	6	6	0	31	10	15	11
Meat meal	10·50	136	6	54	7	4·00	28	0	18	8	0·70	4	2	3	4	76	7	38	4
Fish meal	9·00	117	0	46	9	8·50	59	6	39	8	1·40	8	0	6	5	92	10	46	5
Oats	2·00	26	0	10	5	0·60	4	2	2	9	0·50	3	0	2	5	15	7	7	10
Bran and other offals of wheat	2·50	32	5	13	0	3·60	25	2	16	9	1·45	8	8	6	10	36	7	18	4
Brewers' grains (wet)	0·81	10	6	4	1	0·42	2	11	1	11	0·05	0	4	0	3	6	3	3	2
Clover Hay	2·40	31	2	12	6	0·57	4	0	2	8	1·50	9	0	7	2	22	4	11	2
Dried Grass and Dried Grass Meal	2·50	32	5	13	0	0·75	5	[illegible]	3	7	2·50	15	0	12	0	28	7	14	4
Oat straw	0·50	6	6	2	7	0·24	1	8	1	2	1·00	6	0	4	9	8	6	4	3
Swedes	0·25	3	3	1	4	0·06	0	5	0	3	0·22	1	4	1	1	2	8	1	4
Dried Sugar Beet Pulp	1·30	16	11	6	9	0·25	1	9	1	2	0·40	2	0	1	7	9	6	4	9
Potatoes	0·34	4	5	1	9	0·16	1	[illegible]	0	9	0·60	3	0	2	5	4	11	2	6

modifications of the original tables by Voelcker and Hall were recommended in the light of fresh evidence on the losses involved in dung-making.

The allowance for nitrogen in properly made dung was reduced from 50 to 40 per cent. of the nitrogen consumed in the food; the proportions for phosphoric acid and potash decided upon were 75 per cent. in each case. The allowances were to be halved after one crop has been grown. A further deduction of one quarter from these allowances was recommended when the stock consisted of dairy cows. No difference between food fed upon the land and food fed in yards was to be made.

A recommendation that a Standing Committee should be appointed to revise the tables each year was adopted in Scotland but not in England, and a few figures from the 1954 table of the Scottish Committee, published by the Department of Agriculture for Scotland, are given in Table 64. The valuation is based upon the unit value of N in the cheapest concentrated soluble nitrogenous fertilizer, the unit value of K_2O in muriate of potash and a unit value for P_2O_5 somewhere between that in superphosphate and that in mineral phosphate, allowance being made for appropriate subsidies received. This Committee also decided that the proportions of plant nutrients in the dung should be 40 per cent. for N, 67 per cent. for P_2O_5 and 80 per cent. for K_2O and that the total compensation value should be reduced by 25 per cent. when the food was fed to milking cows.

Of course, no such table can hope to be more than an approximation to the truth ; as has been indicated above, the style of farming must introduce variations special to each case, nor can the table take into account any bad management of land or manure on the part of the farmer. The table assumes ordinary mixed farming and reasonably good management of the dung-heap.

CHAPTER XIV

THE DISTRIBUTION OF FERTILIZERS AND FIELD EXPERIMENTATION

Machines—Field experiments—Sources of error, soil heterogeneity, experimental design, technique, results.

Distribution

IN the use of fertilizers it is very important to get them evenly distributed on the land; nothing is more common in a hay or cereal crop, where a nitrogenous top dressing has been applied, than to see regular waves or lines of a darker green and stronger growth than the bulk of the crop, due to uneven distribution. Irregularities are not so evident with other manures, partly because they are more often sown before or during the final working of the ground and partly because they have not such a striking effect upon the colour and vegetative development of the crop as nitrogen has. Fertilizers can be sown evenly by hand, and this is usually necessary in experimental plots, but materials like ground lime and basic slag are most unpleasant to handle and, for large areas, better and quicker work will always be done by a machine. There are many different types of machine for the purpose, which do good work as long as the fertilizers are fairly dry and friable. Under damp conditions, the moving parts in contact with the manure naturally tend to become clogged up and this interferes with the free flow of the material. The desirable points about a distributor are that it should have few moving parts in contact with the manure, that it should be able to distribute uniformly a wide range of dressings from about 1 cwt. per acre and that, as a result of simple construction, it should be easy to clean. Manufacturers are concerned to see that their fertilizers are delivered in good condition and it rests with the farmer to store them carefully before use and keep to his distributor in good working order.

Machines.—The rising hopper type of machine has long been popular and the principle upon which it works will be gathered from the diagrams, Fig. 5, which show a section through the box *o n*, 8 ft. or 10 ft. 6 in. long. A revolving spindle *m*, running the whole length of the box, is furnished with a series of radial arms which dip into the manure and toss it over the lip of *n*. As the machine travels, the bottom and side *o* of the manure box are raised by the rack and pinion *k* and *l* which are geared to the wheels of the machine; the side *n* remains stationary. The spindle is also geared to the wheels of the machine and the rate at which it revolves and therefore the rate at which the manure is delivered can be varied by changing the gear wheels. After it is tipped over the lip of *n*, the manure falls through a closed channel and can be delivered close to the ground so as to avoid blowing. This machine, therefore, has no parts working in the manure. Delivery can be stopped and started sharply and the rate of sowing can be accurately gauged and, except when the hopper has been freshly filled, distribution is regular.

There are other types in which a feed mechanism works at the bottom of the hopper as shown diagrammatically in Fig. 6. Here the bottom of the hopper A, containing the manure, is formed by a roller or revolving drum B which carries the manure out through the aperture regulated by the adjustable side plate N on to the tray C, from which it is thrown by a brush or revolving spindle with radial arms as in the previous type of machine. The rate of sowing is regulated by the size of the aperture and the speed of the roller. Sometimes there is an agitator in the hopper to stir the manure.

In other cases the manure is scraped through adjustable holes in the bottom of the hopper by a rotating worm or the manure may be carried through adjustable slots by endless chains as illustrated in Fig. 7. The chains come out of the box through a narrow slit dragging with them some of the

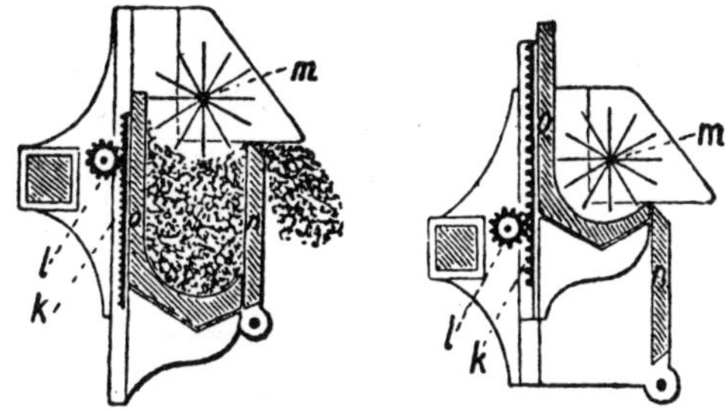

FIG. 5.—DIAGRAMMATIC SECTION OF MANURE DISTRIBUTOR—SEED DRILL TYPE

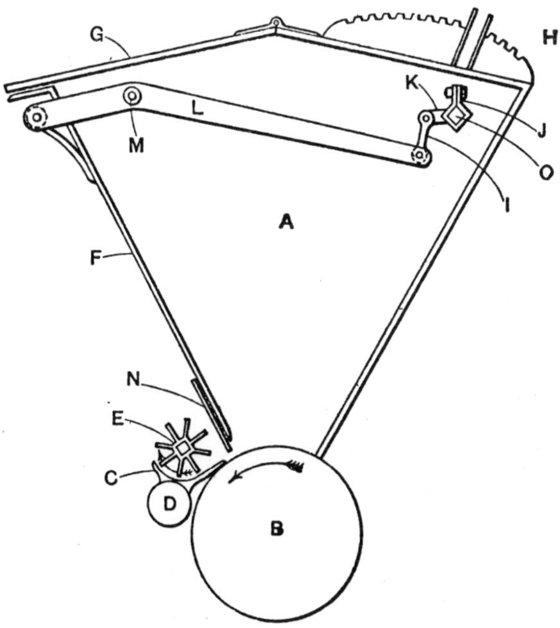

FIG. 6.—DIAGRAMMATIC SECTION OF MANURE DISTRIBUTOR WITH REVOLVING DRUM FEED

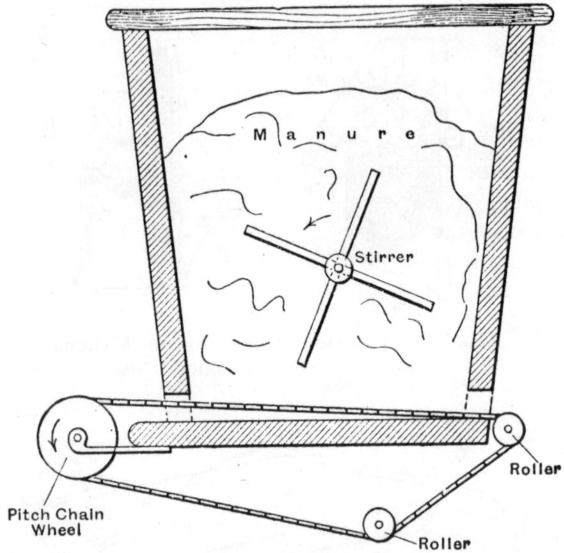

FIG. 7.—DIAGRAMMATIC SECTION OF MANURE DISTRIBUTOR—ENDLESS CHAIN FEED TYPE

FIG 8.—BROADCAST MANURE SOWER WITH REVOLVING DISCS FOR DISTRIBUTION

manure which falls to the ground. The rate of sowing is regulated by the gear wheels which actuate the spindle carrying the pitch chain wheels. A more slowly moving stirrer within the hopper helps the manure moving down to the delivery chains.

Yet another method consists in having a slotted reciprocating plate moving at a controlled speed between two slotted stationary plates which form the bottom of the hopper. Finally there is the system in which finger or star wheels rotate in the bottom of the hopper and push the manure through slots. Both the speed of the wheels and the size of the apertures can be controlled. In recent years there has been a great development in the drilling of seed and fertilizer at the same time by means of combine drills. When the fertilizer is granular, it may be sown with the seed down the same tube, but some machines can sow the two separately. The principle of fertilizer placement has already been discussed in Chapter VI. The modern machine with its star wheel mechanism is very efficient but it is somewhat complicated and demands careful maintenance.

On a different principle are the well-known broadcast distributors of which an example is shown in Fig. 8. Here the manure is carried in a circular hopper and is kept in motion by stirrers. It falls through two adjustable apertures and is intercepted by two horizontal discs with radial ribs, which are kept revolving rapidly by gearing connected with the wheels of the machine. The manure is thus flung in all directions and covers a much wider strip of ground than the track of the machine. Such machines are light and very convenient for some purposes but the distribution is not very uniform. If the manure is a mixture, the heavier particles are thrown further than the light, and very light particles may be carried a long distance in a wind.

Field Experiments

Sources of Error.—The value of any fertilizer on any particular soil can be settled only by experiment, for though it is possible to recognize soil types and to predict their behaviour to some extent, unknown factors often intervene and upset expectations. The proper conduct of field experiments is, therefore, a matter of considerable moment, and it is essential that the degree of accuracy to be expected should be realized before any scheme of experimentation is undertaken. It must not be supposed that experimental error can be got rid of, however carefully the work may be carried out, for in addition to the inherent error in all measurements there is the added difficulty due to the natural variation in soil fertility. The only scientific method of dealing with the error in field experiments is to estimate its magnitude and to avoid drawing any conclusions from results which are not well outside this figure. For example, many trials have shown that, even on apparently uniform ground, the yields of different plots under the same treatment may vary by 10 per cent. This means that if the yields from plots A, B and C are respectively 100, 109 and 91, the conclusion cannot be drawn that treatment B is better than A, or A better than C, because the same variation in the results might have been obtained from three plots all under the same treatment.

An interesting example of this is to be found in the study of the yields from two unmanured grass plots at Rothamsted. Taking a fifty years' average, one plot is 10 per cent. better than the other but nevertheless there were twelve occasions when the first was less than the second. The average superiority of the first could never be assured from the results for one year or even from the average figures for a series of five years, for the error of a single result is $\pm$ 10 per cent. It is unnecessary to quote other figures of the same nature to show that a substantial error is in fact associated with the result from a single plot and must be taken into account in the design of an experiment. Although large single observation plots lying side by side are useful for demonstrating large differences, the numerical results are valueless for comparative purposes because there can be no reliable allowance for experimental errors.

Soil Heterogeneity.—Many factors interfere with experimental control. There may be attacks by birds and insects, incidence of disease and inaccuracies of measurement, but the largest source of error lies in soil heterogeneity. This variation in fertility is systematic rather than random but it cannot be reduced to a simple mathematical formula. Consequently, since the other errors in a carefully conducted experiment are in general much less, it is necessary to reduce this error, and the modern techniques of doing so are due in the first instance to R. A. Fisher. It was he who laid the foundations of the methods of correctly estimating experimental error for small numbers of results and so placed in the hands of the investigator an important tool for checking the accuracy and value of his experimental results. Replication of treatment can diminish the error due to soil heterogeneity and if different treatments can be compared within small blocks (local control) the precision may be increased. But it is important to bear in mind that the design of an experiment must be such that it is possible to make a valid estimate of the error. To achieve this, the treatments must be randomized. Replication, or the laying down of several plots to the same treatment, averages out the experimental error and gives a better picture of the effect of treatment, while the random arrangement of plots means that the effects which are averaged are independent of each other so that an unbiased estimate is obtained. The statistical technique reduces more or less obvious visual observations to figures upon which reliance can be placed and is therefore invaluable for the assessment of results. But it must not be expected to relieve the experimenter of the care and labour required in a serious experiment. It is most important to decide beforehand precisely what questions are to be answered and what design of experiment is most suited to supply adequate data for statistical analysis.

Experimental Design.—The importance of the replication of plots in field work had long been recognized, but the arrangement of the plots was systematic, like ABCDDCBA. Hence, when Fisher (about 1920) devised the various types of experiments to meet the conditions required by statistical analysis, there was considerable criticism by agriculturists because of the complications introduced by randomization. The critics failed to appreciate that there is no method of making a valid estimate of error for many experimental arrangements. In course of time, however, the technique introduced by Fisher has come to be generally accepted as indispensable in all types of biological investigations. There are many volumes and articles devoted to this modern technique of field experimentation and it is beyond the scope of this book to go into details. But attention is directed to the general principles observed in the study of the action of fertilizers in the field.

Needless to say, all plots should be as alike as possible apart from the treatment whose effect is to be measured. In other words, the area selected should be well farmed, but not in a high state of fertility, and by nature uniform with respect to soil and subsoil, drainage, slope and history. It must be remembered that the disturbing effect of farmyard manure or a leguminous crop or the folding of sheep on a portion of a field, will persist for a long time and show up under experimental conditions. The preparation of the ground must be carefully carried out. Any basal dressing applied should be identical for each plot. The plots should lie within a larger area growing the same crop as is under experiment. The size and shape must be the same for all plots and are often decided by local conditions. It will be realized that if the plots are very small they may be grossly affected by patches of abnormal fertility and show great variations, whereas if they are large the error due to trends in soil fertility may become excessive. The optimum size has been found to be about $\frac{1}{40}$ acre for the usual agricultural crops in this country, which means that at least half an acre of ground is required for a comparatively simple experiment.

The shape of the plot is usually a question of convenience. The width, for example, is commonly governed by ease of sowing and harvesting. In the case of a root crop, a useful shape is something like 8 drills wide and 20 yards long; the outside drills can be discarded when the crop is lifted, without introducing a serious sampling error, in order to eliminate edge effects. With cereals, the width is usually a simple multiple of the width of the fertilizer or seed drill used and long narrow plots are very easy to handle at harvest. A clear separation of adjacent plots is obtained by hoeing out one or two outside drills in the early part of the growing season; this also serves to reduce possible error due to the effect of a fertilizer extending beyond the edge of the plot. Each plot then becomes a predetermined length of a definite number of drills and may be cut out without trouble by a binder. When the area is not level, it is of course important that plots under comparison should lie side by side up and down the slope, so that all will be subject to the same probable trend in soil variation. Incidentally, the long narrow plot in a cereal experiment provides the best opportunity of examining the crop during growth.

The methods most commonly used for the arrangement of plots to reduce error and provide a valid estimate of error are (a) Randomized Blocks and (b) the Latin Square. They give the replication and local control previously mentioned. In the former, a number of equal blocks of land are chosen—they may be adjacent or not—and each block is divided so as to contain as many plots as there are treatments. A control is counted as one of the treatments. Each treatment is represented once in each block and only the soil heterogeneity within any one block affects the comparisons. Larger variations in soil fertility over the entire experimental area account for differences in total yields from different blocks and are eliminated from the comparisons by the experimental lay-out. Should it be decided, for example, to have five treatments and six-fold replication, then six blocks are measured off and each is subdivided into five plots, one for each treatment. The positions of the plots must be determined at random for each block. The positions of the blocks are chosen according to site and convenience of working. The blocks may lie side by side or be massed for compactness. This method of lay-out has the great merit of flexibility. In this particular case, the 30 plots provide 29 independent comparisons among the recorded yields, or 29 "degrees of freedom" including 4 for treatments and 5 for blocks. Hence there are 20 "degrees of freedom" to supply a valid estimate of error for the treatment comparisons among the different blocks. A standard error for the results is calculated and differences due to treatments can be subjected to exact tests of significance. That is to say it is possible to tell how frequently a particular difference could occur by chance or what the chances are against such a difference not being due to the treatment in question. In practice, the various treatment yields and the standard error are tabulated as percentages of the mean yield and then information can be obtained from appropriate tables whether the ratio of any difference to the standard error would be exceeded in say 5 per cent. or 1 per cent. of trials. For most field experiments a 30 to 1 chance is accepted as strongly significant and this requires that the difference between treatments should be about 3 times the standard error. For the usual arable crops under careful experiment, the standard error per plot is generally from 5 to 10 per cent. of the yield and, since there are several plots of each treatment, the standard error of the final result is much less—generally about 3 or 4 per cent.

The randomized block method lends itself to the laying down of comparatively simple experiments at a large number of centres; each centre may be regarded as providing a block and replication is obtained through the number of centres. This method is most useful in checking effects under diverse conditions of soil and climate and it may even be possible to group the centres according to certain factors and secure information of value in special cases.

In the Latin Square method, local control is increased by introducing two restrictions in the arrangement of the plots. The replication is the same as the number of treatments and the plots are arranged in a square so that each treatment occurs once in each row of plots and once in each column of plots. Each row and each column may, therefore, be regarded as a randomized block in which the treatments are allocated at random. As the number of plots increases the number of possible arrangements becomes very large. Even with a 4 × 4 square it is a question of selecting one at random out of 576 possible squares. It might be

B	C	D	A
D	B	A	C
A	D	C	B
C	A	B	D

and is really only a conventional square for the plots are usually oblong.

Soil differences can be eliminated in two directions, for both rows and columns are replicates of each other. This method generally provides more accurate comparison than the method of randomized blocks and is quite convenient when fertilizers are to be applied by hand as is usual with small plots. But when the number of treatments exceeds 6 or 7 it becomes less practicable than the randomized blocks method.

When the experiment includes a large number of treatments it is quite common to subdivide the plots so that combinations of treatments can be tested. This is a very useful procedure when two series of treatments are involved such as in an experiment with, for example, two rates of dressing of both phosphatic and potassic fertilizers. Here we have nine different combinations in all. The results will provide information not only on the response of the crop to the different dressings alone, but also on the interaction between phosphate and potassium. Such a lay-out appears to be complicated, but in actual fact is not difficult to deal with and the arrangement not only gives greater replication for straight comparisons but shows how the effects of one treatment are influenced by the other. A further development has been the design in which each block does not include all treatments but where one set of treatments is "confounded" with another. It is mentioned simply to demonstrate the extraordinary flexibility of the modern technique of field experimentation.

Technique.—With respect to the precautions necessary in carrying out an experiment, several points immediately come up for consideration. Great care should be exercised in measuring and marking the plots and, when the experiment is to be continued for more than one season, some method of permanent identification of the plots is necessary. This may be done by driving stout posts to below cultivation depth at the corners of the experimental area; if a loose wire is attached to the top of each post it assists when the time comes for locating the corners by probing. The positions of the corners are fixed by their distances from permanent stakes at the edge of the field. When the plots are continued year after year it is necessary to watch the method of ploughing. If a turn-wrest plough is used, the mould board should be set to throw the furrows opposite ways in alternate ploughings, otherwise the manured soil will gradually be displaced sideways. If plots are ploughed in lands, the furrows should be alternately gathered to and cast away from the middle of the plot.

An aid to the good distribution of a manure is to portion it out to different sections of the plot or to mix it with soil to increase the bulk so that the plot can be crossed in different directions. Wind often presents a serious problem in some districts when the fertilizer is finely divided and the second method is then the better if operations cannot be delayed until calm conditions prevail.

Rather more care than usual should be given to the singling of root crops so as to obtain a uniformly set-out plant. It may not be desirable to fill up gaps and misses by transplanting because this conceals the possible effect of a manure on the germination and development of the seedling. The number of plants on a plot should be counted before harvesting as the figure often affords a means of criticizing the weights and of estimating the influence of the manure upon the constitution of the plant. The roots should be carefully topped and both tops and roots weighed, preferably in wire baskets.

Cereal crops can be cut either by scythe or machine. If the plots are long and the divisions between them clear, an ordinary binder can be successfully employed; it would be a great advantage if a combine-harvester type of machine were available to facilitate the exacting nature of the work. The produce should be gathered on to the plot from which it has been cut and clearly marked. Sometimes threshing can be carried out on the field, but it is often necessary to carry the unthreshed sheaves to a rick or barn. Loss of grain in carting may be prevented by inserting each sheaf head-first into a sack. To keep the produce of each plot separate until threshing time, a small "hut" may be built for each plot; if this is inconvenient, squares of fabric, sufficiently open to allow ventilation but not the passage of shed corn, may be used to keep plot layers separate in the stack. When a small threshing machine is not available, the ordinary modern travelling mill will serve the purpose. But the screens should be removed at the end of each run and a little straw put through the machine to work out all the grain of which the last pint or so is extracted by hand from the hopper. The grain should be measured out bushel by bushel and every bushel weighed and recorded; the tail corn should be weighed as a whole; the straw and cavings should also be weighed.

Complete records should be made of all measurements. It is inadvisable to make additions or subtractions mentally on the field; there should be a place on the record for the entry of all figures as they are obtained. Sometimes at harvest it is necessary to discard a whole plot for some reason—a mistake in treatment or local damage by an animal—and fortunately it is possible to make an estimate of the missing yield by statistical analysis and so save the symmetry of the experiment. An experiment demands the greatest possible measure of care and patience. Failure to take all precautions simply means a waste of effort.

Results.—Finally, it may be that the object of the experiment requires more than mere yield figures. In the case of turnips it is usually desirable to determine the production of dry matter, or the composition of the ash may be wanted to measure the recovery of some element from the fertilizer applied. This involves the collection of cores from a definite number of roots (usually 50 or 100) selected at random from each plot. In sugar-beet experiments, the percentage of sugar is required; for cereals the content of protein in the grain is commonly determined. When the crop is grass for hay, then dry-matter figures are essential to obtain a correct picture of the yield under different treatments. This means that a proper sample, representative of the herbage, must be taken from each plot for laboratory examination. It is extremely difficult to assess the yield or quality of pasture, for so many diverse factors are involved in a measure of productivity. Yields of dry matter and of various organic and inorganic constituents are useful guides, but such questions as palatableness and actual nutritive value remain to be determined by direct feeding to stock. But measurements of live weight increases cannot be accurate on small plots, and estimates of carrying capacity are often subject to personal opinion. Some grassland experiments have been extended to the evaluation of the animal carcasses at the slaughter-house and there is no question that very large and significant differences in the returns from different cultural and manurial treatments frequently emerge; but it is quite probable that smaller although financially important differences are masked by the large experimental errors.

There should be no need to emphasize the necessity of repeating an experiment at several centres and in several years to overcome the variations which are bound to exist due to different soil types and to different seasons. This demands an immense amount of labour and can be satisfactorily carried out only by co-operation among several workers interested in the same problems. The results so obtained are of much more general application and much more valuable to the agricultural community than the single experiment however well it may have been conducted.

It is sometimes asked how far very small plots, cultivated with all the care and attention given by a good gardener to his plants, can be made to serve for experimental work on fertilizers. They are quite satisfactory for demonstration purposes and on occasion have shown quite low experimental error in quantitative work, but the applicability of the results to farm practice is of necessity doubtful. In pot work the artificiality of the conditions is increased and deductions are naturally to be drawn with great caution. But pot experiments are of great value to the investigator in the preliminary stages of an inquiry before the applications to practice are considered. They enable him to control some factors which are not under control in the field and to study different soils under the same conditions at one centre.

It follows from what has been said about the care with which experiments must be carried out, and the error inherent in the results, that they are not to be lightly undertaken by the ordinary busy farmer. Single-handed it takes too long to arrive at reliable conclusions and the farmer's experimenting is best done as part of a co-operative trial designed to establish the characteristics of the soils in his area. The common habit of testing a fertilizer by differential treatments across a field may often lead to erroneous conclusions. Nitrogenous fertilizers in particular give marked responses in colour and vegetative appearance which are deceptive, whilst basic slag may bring about a remarkable change in the botanical composition of a pasture. Generally speaking, the effects of fertilizers can be measured only by yield. Numerous results have shown that it is difficult, even for the experienced observer, to detect by eye the difference between adjoining plots whose yields actually differ by 20 per cent. That is a fact which is commonly ignored or disbelieved, and justifies the need for work which can give reliable results for differences of much less than 20 per cent. but nevertheless of considerable monetary importance.

TABLES FOR THE CONVERSION OF NITROGEN INTO AMMONIA AND "PHOSPHORIC ACID" (P_2O_5) INTO TRI-CALCIUM PHOSPHATE.

Ammonia.		Nitrogen.	Nitrogen.		Ammonia.
1	=	0·8235	1	=	1·214
2	=	1·647	2	=	2·429
3	=	2·471	3	=	3·643
4	=	3·294	4	=	4·857
5	=	4·118	5	=	6·071
6	=	4·941	6	=	7·286
7	=	5·765	7	=	8·500
8	=	6·588	8	=	9·714
9	=	7·412	9	=	10·93

Tri-calcium Phosphate.		"Phosphoric Acid."	"Phosphoric Acid."		Tri-calcium Phosphate.
1	=	0·4576	1	=	2·185
2	=	0·9152	2	=	4·370
3	=	1·373	3	=	6·556
4	=	1·831	4	=	8·741
5	=	2·288	5	=	10·926
6	=	2·746	6	=	13·111
7	=	3·302	7	=	15·297
8	=	3·661	8	=	17·482
9	=	4·119	9	=	19·667

Example.—Reduce 6·43 per cent. Ammonia and 35·21 per cent. Tri-calcium Phosphate to Nitrogen and "Phosphoric Acid":

Ammonia		Nitrogen.	Tri-calcium Phosphate.		"Phosphoric Acid."
6	=	4·941	30	=	13·73
0·4	=	0·329	5	=	2·288
0·03	=	0·025	0·2	=	0·091
			0·01	=	0·005
		5·295			16·114

POTASSIUM AND POTASH.

100 parts of Potassium (K) are equivalent to 120·5 parts Potash (K_2O).
83 ,, ,, ,, 100 ,, ,,

Potassium sulphate (K_2SO_4) contains 44·83 per cent. K.
(equivalent to 54·05 per cent. K_2O).
Potassium chloride (KCl) contains 52·44 per cent. K.
(equivalent to 63·18 per cent K_2O).

INDEX